10/27/94

D1033926

NONLINEAR REGRESSION MODELING

STATISTICS: Textbooks and Monographs

A SERIES EDITED BY

D. B. OWEN, Coordinating Editor
Department of Statistics
Southern Methodist University
Dallas, Texas

OTHER VOLUMES IN PREPARATION

NONLINEAR REGRESSION MODELING

A Unified Practical Approach

David A. Ratkowsky

CSIRO Division of Mathematics and Statistics
Tasmanian Regional Laboratory
Hobart, Tasmania

MARCEL DEKKER, INC.   New York and Basel

Library of Congress Cataloging in Publication Data

Ratkowsky, David A., [date]
  Nonlinear regression modeling.

  (Statistics, textbooks and monographs ; v. 48)
  Bibliography: p.
  Includes index.
  1. Regression analysis. 2. Parameter estimation.
I. Title. II. Series.
QA278.2.R37 1983   519.5'36   83-6599
ISBN 0-8247-1907-7

MARCEL DEKKER, INC.
270 Madison Avenue, New York, New York 10016

Current printing (last digit):
10 9 8 7 6 5 4 3

PRINTED IN THE UNITED STATES OF AMERICA

Preface

Linear regression models are ones in which the parameters appear
linearly, whereas nonlinear regression models have at least one
parameter appearing nonlinearly. This book is devoted to a study of
the behavior in estimation of nonlinear regression models, a subject
which is still very much in its infancy. Despite the large number of
users of nonlinear regression models and a vast literature on methods
for finding the least-squares estimators of the parameters of nonlinear
models, it is only in the last two decades that any serious attempt
has been made to study the properties of those estimators. Even so,
in the space of twenty years the number of papers on this subject has
been remarkably few. Nonlinear regression models differ greatly in
their estimation properties from linear regression models in that,
given the usual assumption of an independent and identically
distributed normal stochastic term, linear models give rise to
unbiased, normally distributed, minimum variance estimators, whereas
nonlinear regression models tend generally to do so only as the sample
size becomes very large. The purpose of this book is to explore the
properties of the estimators in problems having sample sizes similar
to those actually obtained in practice by scientists in agricultural
research, biology, engineering, and other applied disciplines.

Modelers probably have one of three purposes in mind when they
wish to fit a nonlinear regression model to a set of data. They may

be interested in (1) obtaining a "good fit" to the data merely for the purposes of representation, (2) predicting response values Y for given fixed values of the regressor variable X, and (3) making inferences based upon interpretation of the parameter estimates. This book is concerned with the second and third purposes, the first purpose being outside its scope. In nonlinear regression models the predicted values of Y will be biased, the extent of the bias depending upon a quantity known as the "intrinsic nonlinearity" of the model/data set combination. Interpretation of parameter estimates also involves an additional quantity known as the "parameter-effects" nonlinearity. Even if the intrinsic nonlinearity is acceptably low, high parameter-effects nonlinearity may make it difficult to obtain convergence to the least-squares estimates of the parameters using standard algorithms such as the Gauss-Newton method. It is argued throughout this book that, provided the intrinsic nonlinearity is acceptably low, it is desirable to seek models with low parameter-effects nonlinearity and that such models may often be obtained by appropriate reparameterizations. A suitably reparameterized model will behave like a linear model and the Gauss-Newton method will converge rapidly to the least-squares estimates, rendering unnecessary the various complex methods that have been developed for obtaining parameter estimates.

It is assumed that the reader of this book has a background in statistical methods and computing and has a knowledge of the theory and practice of linear regression analysis at a level such as that presented in standard texts such as Draper and Smith (1981) and Gunst and Mason (1980). Scientists in various practical disciplines who have access to an applied statistician or biometrician should also be able to use the methods espoused in this book.

There are many individuals to whom I owe a debt of gratitude. Peter Gillis, formerly Chief Biometrician with the Tasmanian Department of Agriculture, has been a close collaborator in this work from the outset. The material in Chapters 3 and 7 could not have been written without his active interest and cooperation. I have benefitted greatly from the numerous hours of discussion that I have had over a period of 12 years with Glen McPherson, Lecturer in Statistics at the University of Tasmania, who was also the first to suggest that I write this book. I received much assistance from several colleagues in the CSIRO Division of Mathematics and Statistics stationed in other Australian cities. They have been most generous in reading the various chapters of this book and making detailed suggestions for improvements. Among these colleagues I wish particularly to thank Dr. Norm Campbell, Perth, Western Australia. Also, Dr. Doug Shaw, Sydney, New South Wales, and Dr. Richard Tweedie, formerly of Melbourne, Victoria, read some of the chapters.

The comments of Dr. Phil West of the CSIRO Divison of Forest Research, Hobart, as a reader interested in applying the methodology but who has not had any formal training in statistics , have been most helpful in providing me with some insight into how a non-statistician views the problems of nonlinear regression modeling. All illustrations in the book were made by Tony Quarmby, CSIRO Division of Food Research, Hobart. Thanks also go to Jill Adamski, Librarian of the CSIRO Tasmanian Regional Laboratory, for obtaining articles, books, and references. The computer programs in the Appendix were written by Rob Lowry, CSIRO Division of Mathematics and Statistics, Hobart, who also assisted with computation throughout all stages of the preparation of this book. He has also contributed substantially by proposing the asymmetry measure of bias described in Section 2.9. The computer code for the curvature measures of nonlinearity was adapted from a program of Dr. Doug Bates, Madison, Wisconsin, for whose help I am grateful. Cate Lowry typed the manuscript and the final camera-ready copy. Without the assistance of all the above-mentioned people this book may never have been written. I take sole responsibility for any mistakes herein.

David A. Ratkowsky

Contents

1

Introduction to Regression Models

1.1 Linear Regression Models

The term "linear regression model" is sometimes used to mean two different things. The first usage refers to a straight-line relationship between two variables, and the second refers to a model in which the *parameters*, the quantities to be estimated, appear linearly. It is the latter usage which will be followed throughout this book. For example, the following are linear regression models:

$$Y = \alpha + \beta X + \varepsilon \tag{1.1}$$

$$Y = \alpha + \beta X + \gamma X^2 + \varepsilon \tag{1.2}$$

$$Y = \alpha + \beta_1 X_1 + \beta_2 X_2 + \varepsilon \tag{1.3}$$

where the Greek letters α, β, β_1, β_2, and γ represent unknown parameters believed to be constant for a given model/data set combination, whereas X, X_1, and X_2 are variables (often called regressor, predictor, or independent variables) which may represent experimental settings, predetermined conditions, or uncontrolled observed values assumed to be measured without error. In general they are not *random* variables. The response variable Y, also called the dependent variable, deviates from the expected (i.e., mean) value given by the regression line by an amount ε, which is an unobservable random "error" term whose values are unknown but which is assumed to have a mean value of zero.

Model (1.1) may be seen to be the "straight-line regression"

1

model, where α is the intercept and β is the slope. Model (1.2) is
of higher order in X and is the equation of a parabola; although the
relationship between Y and X is nonlinear, model (1.2) is still said
to be linear, as the parameters α, β, and γ appear linearly. Model
(1.3) is the most simple example of a "multiple-variable linear
regression" model, there being two regressor variables, X_1 and X_2.

Regression models such as (1.1)-(1.3) may be written in
mathematical notation in a variety of ways. Considering only model
(1.1), the following alternative forms may be encountered:

$$Y_t = \alpha + \beta X_t + \varepsilon_t \qquad (1.4)$$

$$E(Y) = \alpha + \beta X \qquad (1.5)$$

$$E(Y|X) = \alpha + \beta X \qquad (1.6)$$

The form (1.4) reminds the user that the data set consists of n available
responses Y_t (t = 1, 2, ..., n) measured at corresponding values of
the independent variable X_t (t = 1, 2, ..., n). These n pairs of
observations may be written (X_1, Y_1), (X_2, Y_2), ..., (X_n, Y_n). For
each pair (X_t, Y_t), there is a "perturbation" ε_t which causes Y_t to
deviate from its expected value $\alpha + \beta X_t$, given that $E(\varepsilon) = 0$. The
left-hand sides of (1.5) and (1.6) are alternative ways of expressing
the belief implicit in the use of model (1.1) that the expected value
of Y for a specified X is $\alpha + \beta X$. One may also write the model simply
as

$$Y = \alpha + \beta X \qquad (1.7)$$

to specify only its deterministic component.

The basis for estimating the unknown parameters in all models used
in this book, whether they are linear or nonlinear models, will be the
criterion of *least squares* (LS). Other criteria are available, but
it will not be within the scope of the present work to explore them.
It may suffice to say that least squares has some optimum properties
when certain conditions are met. For a detailed account of these
optimum properties, see Malinvaud (1970, Chaps. 3 and 5). In this book
we will further assume that the ε_t are independent and identically

distributed normal (iidN) random variables with mean zero and finite variance σ^2, in which case the LS estimators of the unknown parameters in linear models are also the *maximum likelihood* (ML) estimators of the parameters in the model. They are unbiased, and among the class of linear regular unbiased estimators, have the property of being minimum variance estimators. Given that the assumptions noted above are satisfied, the criterion of least squares thus provides the best available estimates in practice.

Let us consider a very simple data set (Table 1.1) which will be useful in illustrating the difference between linear and nonlinear regression models. As there are only two pairs of observations, n = 2, the pairs (X_t, Y_t) being $(2, 2.5)$ and $(3, 10.0)$. These data are displayed in Fig. 1.1 as a conventional plot of Y_t versus X_t for each value of t. Suppose that there is reason to believe, from knowledge about the system or process from which these data were obtained, that the simple model of "straight-line regression through the origin" applies:

$$Y_t = \beta X_t + \varepsilon_t \tag{1.8}$$

The LS estimator of β is obtained by minimizing the sum of squares of deviations of the observed Y_t from the assumed true model, that is, by minimizing

$$S(\beta) = \sum_{t=1}^{n} (Y_t - \beta X_t)^2 \tag{1.9}$$

Subsequently, the summation limits will be omitted from $\sum$ for simplicity. Writing S in place of $S(\beta)$ to simplify the notation, the

Table 1.1 Illustrative Data Set

Observation number	X	Y
1	2	2.5
2	3	10.0

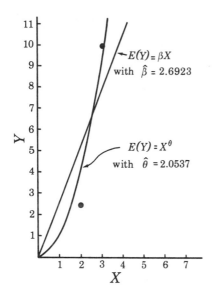

Fig. 1.1 Data observations from Table 1.1 with fitted models (1.8) and (1.11).

minimum value of S may be obtained by differentiating (1.9) with respect to β, setting the derivative equal to zero, and solving for β, whose solution is denoted $\hat{\beta}$ to indicate that it is the LS estimator of β. Hence

$$\frac{\partial S}{\partial \beta} = -2 \sum (Y_t - \beta X_t) X_t = 0$$

which leads to the following:

$$\hat{\beta} = \frac{\sum X_t Y_t}{\sum X_t^2} \qquad (1.10)$$

As the X_t have been assumed to be without error and are not random variables, $\hat{\beta}$ is a linear combination of the random variables Y_t. If the Y_t are assumed to be normally distributed about their mean values βX_t with finite (albeit unknown) variance σ^2 (i.e., the variance of

ε_t), it follows that $\hat{\beta}$ is also normally distributed. Furthermore, the expectation of $\hat{\beta}$ is β, which means that $\hat{\beta}$ is unbiased for β, and the variance of $\hat{\beta}$, which is

$$\text{Var } (\hat{\beta}) = \frac{\sigma^2}{\sum x_t^2}$$

is the minimum possible variance of any linear unbiased estimator for β. For further information on linear regression models and their use, readers are referred to practical textbooks such as those by Draper and Smith (1981) and Gunst and Mason (1980).

1.2 Nonlinear Regression Models

A nonlinear regression model is one in which the parameters appear nonlinearly, for example

$$Y_t = X_t^\theta + \varepsilon_t \tag{1.11}$$

where θ is the parameter to be estimated. In a similar fashion to linear models, one may use least squares to estimate θ by minimizing

$$S(\theta) = \sum (Y_t - X_t^\theta)^2 \tag{1.12}$$

Writing S in place of $S(\theta)$ to simplify the notation, the minimum of S may be obtained by differentiating (1.12) with respect to θ, setting the derivative equal to zero, as follows:

$$\frac{\partial S}{\partial \theta} = -2 \sum (Y_t - X_t^\theta)(\log X_t)X_t^\theta = 0$$

and attempting to solve for θ, the solution to which is denoted $\hat{\theta}$. However, this does not lead to an explicit expression for $\hat{\theta}$. Instead, the resulting rearranged equation

$$\sum Y_t(\log X_t)X_t^{\hat{\theta}} = \sum (\log X_t)X_t^{2\hat{\theta}} \tag{1.13}$$

can yield the LS estimate $\hat{\theta}$ only by an iterative procedure starting from some assumed value of $\hat{\theta}$.

For the data set in Table 1.1, where the pairs (X_t, Y_t) were (2, 2.5) and (3, 10), $\hat{\theta}$ in model (1.11) can be determined to be 2.0537 (to four-decimal-place accuracy); this fitted model is displayed in Fig. 1.1. An important point is that the LS estimator $\hat{\theta}$ of θ in (1.11) does not have properties possessed by $\hat{\beta}$ in the linear model (1.8). Only "asymptotically," that is, as the sample size increases to infinity, do the properties of $\hat{\theta}$ approach the properties of $\hat{\beta}$. For a detailed account of this asymptotic or large-sample behavior of the LS estimator in nonlinear regression models, the reader is referred to Jennrich (1969) and Malinvaud (1970, Chap. 9). For finite samples, the general statement may be made that even though Y_t may be normally distributed about its mean X_t^θ with some finite unknown variance σ^2 for all t, t = 1, 2, ..., n (so that the LS estimator $\hat{\theta}$ is also the ML estimator of θ), $\hat{\theta}$ is *not* a linear combination of the Y_t and hence in general is *not* normally distributed, *nor* is it unbiased for θ, *nor* is it a minimum variance estimator. Thus, unlike a LS estimator of a parameter in a linear model, a LS estimator of a parameter in a nonlinear model has essentially unknown properties for finite sample sizes. Nevertheless, as the sample size n increases, one may predict how $\hat{\theta}$ will tend to behave provided that ε_t is iidN (in which case the LS estimator is also the ML estimator). Under these circumstances, asymptotic theory tells us that the ML estimator becomes more and more unbiased, more and more normally distributed, and approaches a minimum possible variance (called the minimum variance bound) more and more closely as the sample size becomes larger (see Kendall and Stuart, 1967, Chap. 17). In general, however, it is not possible to present any guidelines as to how large the sample size must be for these so-called asymptotic properties to be closely approximated. It will be seen in Chapters 3-6 that there are some nonlinear models for which the asymptotic properties are a good approximation even for very small sample sizes, whereas there are other nonlinear models where the asymptotic properties are poorly approximated even for what must be considered to be very large samples in practical terms.

1.3 Geometrical Representation of Nonlinear Regression Models

A better understanding of the way in which nonlinear models differ from linear models may be obtained using an alternative way of depicting the data. This approach has also been used by Draper and Smith (1981, Secs. 10.5 and 10.6). We employ the data set of Table 1.1, consisting of the two pairs of observations (X_t, Y_t), viz., (2, 2.5) and (3, 10.0). Instead of graphing the data in the traditional way as in Fig. 1.1, by plotting Y_t versus X_t, the alternative approach involves using a separate coordinate axis for each pair of observations. In this case n = 2, so two coordinate axes are needed, as shown in Fig. 1.2, to display the sample in n-dimensional "sample space." As Y_t are Y_1 = 2.5 and Y_2 = 10, respectively, a "Y-vector" is drawn from the origin to the vertex (2.5, 10).

Considering now the nonlinear model (1.11), rewritten as

$$E(Y) = X^\theta \tag{1.14}$$

it is possible to draw (Fig. 1.2a) the *solution locus* for this model, this being the locus of points having the form (X_1^θ, X_2^θ), in this case $(2^\theta, 3^\theta)$, representing all possible values of θ, or "solutions" to the model (1.14). The LS estimate $\hat\theta$ is that point on the solution locus *nearest* Y. This may be obtained geometrically by dropping a perpendicular from the tip of the arrowhead at Y to lines tangent to the solution locus at various points along the locus and ascertaining the perpendicular distance of minimum length. This minimum perpendicular distance is shown in Fig. 1.2a and corresponds to $\hat\theta$ = 2.0537, which, in terms of coordinates 1 and 2, is $(2^{2.0537}, 3^{2.0537})$ or (4.152, 9.547). These are, in fact, the "fitted" or "predicted" values $\hat Y$ corresponding to X_1 = 2 and X_2 = 3, respectively (see Fig. 1.1).

The shape of the solution locus in the vicinity of $\hat\theta$, and the spacing of values of constant $\Delta\theta$ on the solution locus in the vicinity of $\hat\theta$, may be used as measures of the degree to which a nonlinear model differs from a linear model. This possibility follows from the fact

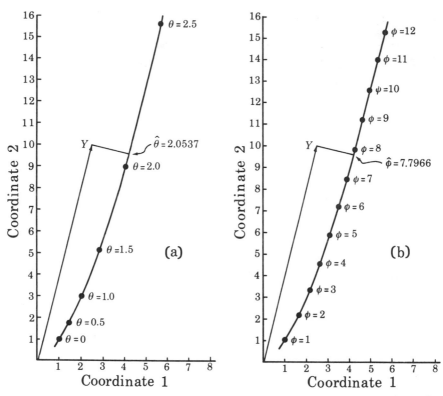

Fig. 1.2 Sample space representation of data: (a) model $E(Y) = X^\theta$; (b) model $E(Y) = X^{\log\phi}$.

that, in a linear model, the solution locus is linear, being a straight line for a one-parameter model such as we have been considering, and that points on the linear solution locus corresponding to equal increments of θ are themselves equally spaced. Referring to Fig. 1.2a, the solution locus for (1.14) is seen to be slightly curved in the vicinity of $\hat\theta$. The extent of this curvature has been called both by Beale (1960) and by Bates and Watts (1980) the *intrinsic nonlinearity* and becomes defined the moment a model/data set combination has been specified. The name "intrinsic" is very suitable, as the intrinsic nonlinearity cannot be altered by reparameterization. A quantitative measure of this intrinsic component of nonlinearity will be presented in Chapter 2, but it is obvious from Fig. 1.2a that it is not large for model (1.14) in combination with the data of Table 1.1, as the solution locus is only slightly curved near $\hat{\theta}$.

The other way in which nonlinear models differ from linear models involves the position on the solution locus of values of θ having equal increments of θ, that is, $\Delta\theta$. In Fig. 1.2a, points for $\theta = 0$, $\theta = 0.5$, $\theta = 1$, $\theta = 1.5$, $\theta = 2$, and $\theta = 2.5$, corresponding to $\Delta\theta = 0.5$, are not equally spaced as they would be in a linear model, but become increasingly more distant as θ increases. The extent to which the spacing is unequal in a nonlinear model defines the extent of this second component of nonlinearity. This component has been termed the *parameter-effects* nonlinearity by Bates and Watts (1980), because its extent is determined by the way in which the parameters appear in the model. In Chapter 2 we will introduce a quantitative measure of parameter-effects nonlinearity derived by Bates and Watts (1980), which is based on the spacing of $\Delta\theta$ when θ is projected onto the tangent plane to the solution locus at $\hat{\theta}$, rather than on the spacing of $\Delta\theta$ on the solution locus itself. For the moment, we confine ourselves to the qualitative remark that parameter-effects nonlinearity may often be reduced, sometimes drastically, by suitable reparameterization. For example, if we reparameterize model (1.14) as

$$E(Y) = X^{\log\phi} \tag{1.15}$$

then the LS estimate of ϕ for the data of Table 1.1 is $\hat{\phi} = \exp(\hat{\theta}) = 7.7966$. Figure 1.2b shows values of ϕ corresponding to $\Delta\phi = 1$ on the solution locus and it may be seen that the spacing of these equal increments of ϕ is almost equal throughout its range. Hence model (1.15), which is a reparameterization of model (1.14) through the function

$$\phi = \exp(\theta) \tag{1.16}$$

may be expected to exhibit a closer approach to linearity than model (1.14). That this is in fact the case will be quantified by various measures and tests to be presented in Chapter 2.

1.4 The Concept of Nonlinear Estimation Behavior of Nonlinear Regression Models

Although the geometrical representation of nonlinear regression models
employed in Sec. 1.3 is restricted in practice to sample size $n \leq 3$
and number of parameters $p \leq 2$, the conceptual principle is valid for
higher dimensions as well. A linear regression model has a linear
solution locus, which means a straight line for $p = 1$, a plane for
$p = 2$, a hyperplane for $p \geq 3$, and an intrinsic nonlinearity of zero
irrespective of the data with which the model is in combination. In
addition, lines on the solution locus, which may be called "parameter
lines," representing constant values of θ_i, where θ_i is an element of
the parameter vector $\underset{\sim}{\theta}$ ($i = 1, 2, \ldots, p$), are straight and parallel
and are equally spaced for equal increments of θ_i. For a nonlinear
regression model the situation is different. The solution locus is
a curved surface, and the parameter lines on the solution locus (or
the projections of these lines onto the tangent plane to the solution
locus at $\hat{\underset{\sim}{\theta}}$, as used by Bates and Watts, 1980) are, in general, neither
straight, parallel, nor equispaced. Thus, measures such as those
proposed by Bates and Watts (1980), which quantify the curvature of
the parameter lines and their lack of parallelism and equispacedness,
can provide the modeler with an effective approach to what may be
described as the *nonlinear behavior* of a nonlinear model.

There is, however, a second approach to assessing the nonlinear
behavior. This follows from the properties of the LS estimators of
the parameters of linear models that were discussed in Sec. 1.1. Given
the assumption of iidN error, the LS estimator $\hat{\underset{\sim}{\theta}}$ of $\underset{\sim}{\theta}$ in linear models
is unbiased, normally distributed, and achieves the minimum variance
bound; for nonlinear models, the LS estimator has these properties only
asymptotically. However, we can speak of the nonlinear behavior of
a nonlinear model in terms of the *sampling properties* of the LS
estimator. If the LS estimator of a parameter of a nonlinear model
is only slightly biased with a distribution close to that of a normal
distribution and with a variance only slightly in excess of the minimum

variance bound, it seems reasonable to speak of the estimator as behaving *close to linear* since its properties are close to what one expects from a linear model. If, on the other hand, the LS estimator is badly biased, with a distribution far from normal and a variance greatly in excess of the minimum variance bound, it is clear that the nonlinear model is, in behavior, far from a linear model.

It will be one of the main purposes of this book to explore these two different approaches to assessing nonlinearity in nonlinear models and to see how closely they agree when they are applied to practical models using real data. It will be taken as self-evident that, given a choice, a close-to-linear nonlinear model is to be preferred to one whose behavior is far from linear. This follows from the desirable properties of linear models, some of which were discussed in Sec. 1.1. Other desirable properties of close-to-linear models, as the reader will see in subsequent chapters, include the ease of obtaining the LS estimates and their relatively straightforward interpretation. Other benefits are that predicted values of the response variable Y will be almost unbiased and that joint confidence regions for the parameters will be close to the ellipsoidal shape characteristic of linear models.

Exercises

1.1 Suppose that the data set for the example in Sec. 1.1 had consisted of the pairs (2, 2.5) and (3, 1.0) instead of (2, 2.5) and (3, 10.0) as in Table 1.1. Verify that $\hat{\theta}$, the LS estimate of θ, for model (1.14) in combination with these data, would have been approximately 0.5. For this case, would you expect (a) the intrinsic nonlinearity, and (b) the parameter-effects nonlinearity, to be greater or less than for the original data pairs in Table 1.1?

1.2 Consider the linear model $E(Y) = \beta X$ in combination with the data of Table 1.1. Draw the solution locus for this model/data set combination on a graph in sample space (similar to that shown in Fig. 1.2). Display the points on the solution locus corresponding to β = 0, 1, 2, 3, 4, respectively, for which $\Delta\beta$ = 1. What can be said about (a) the shape of the solution locus, and (b) the spacing of these points, compared with those for the nonlinear model $E(Y) = X^\theta$?

1.3   The position of the solution locus in Fig. 1.2b is identical with that in Fig. 1.2a, implying that reparameterization of a model leaves the position of the solution locus unaltered. By comparing model (1.15) with model (1.14), show that this must be true. Show that a predicted value of Y, for a given X value, is the same whether model (1.15) or (1.14) is used.

2

Assessing Nonlinearity in Nonlinear Regression Models

2.1 Introduction

In Chapter 1 it was noted that nonlinear regression models differed in general from linear regression models in that the LS estimators of the parameters are biased, non-normally distributed, and have variances exceeding the minimum possible variance. The extent of the bias, non-normality, and "excess" variance differs widely from model to model. It is actually more appropriate to speak of a "model/data set" combination, rather than of a "model," since a specific set of observed data in conjunction with a specified model determines its behavior. For simplicity, however, the word "model" will be used when it is clear that it relates to a specific data set. As the sample size increases, the behavior of the LS estimators becomes more linear, but until recently, there were no readily applicable methods for quantifying the nonlinear behavior.

Beale (1960) made the first serious attempt to measure nonlinearity, but his measures tend to predict that a model will behave linearly even when considerable nonlinearity is present (Guttman and Meeter, 1965). Box (1971) presented a formula for estimating the bias in the LS estimators, and Gillis and Ratkowsky (1978), using simulation studies, found that this formula not only predicted bias to the correct order of magnitude in yield-density models but also gave a good indication of the extent of nonlinear behavior of the model; their simulation studies showed that the extent of the bias was closely associated with the extent of the non-normal behavior of the estimator. Bates and Watts (1980) developed new measures of nonlinearity based on the geometric concept of curvature. They showed that the nonlinearity of a model can be separated into two components: (1) an "intrinsic" nonlinearity associated with the curvature of the solution locus, and (2) a "parameter-effects" nonlinearity associated with the fact that the projections of the parameter lines on the tangent plane to the solution locus are, in general, neither straight, parallel, nor equispaced. They demonstrated the relationship between their measures and those of Beale (1960), and explained why Beale's measures generally tend to underestimate the true nonlinearity. In addition, they showed that the bias measure of Box (1971) is closely related to their measure of parameter-effects nonlinearity. In the 24 model/data set combinations that they studied, the parameter-effects nonlinearity was always greater, often very much greater, than the intrinsic nonlinearity. This helps explain why Gillis and Ratkowsky (1978) found that the bias measure of Box (1971), which is associated only with parameter-effects nonlinearity, was closely correlated with the properties of the LS estimator as revealed by simulation studies.

Throughout this book, the expressed goal will be, for a given data set, to find a model which comes close to behaving "linearly." This means finding a model for which the LS estimators of its parameters are almost unbiased, normally distributed, and whose variances are close to the minimum variance. Such a model should have both a low intrinsic nonlinearity and a low parameter-effects nonlinearity. A negligible intrinsic nonlinearity will mean negligible bias in predicted values of Y (see Sec. 9.2); if, simultaneously, the parameter-effects nonlinearity is also negligible, the more valid will be statistical tests whose justification rests on the assumption of linearity (see Sec. 9.3). Sections 2.4–2.9 describe in detail several measures or procedures for assessing nonlinearity in nonlinear models. First, two important questions are considered: that of obtaining the LS estimates of the parameters in the model (Sec. 2.2), and that of deciding whether the model provides a good fit to the data (Sec. 2.3).

2.2 Obtaining the LS Estimates of the Parameters

There is a wealth of literature on how to determine the LS estimates
of the parameters once a nonlinear model has been specified and a set
of data obtained (e.g., see the review paper by Chambers, 1973). This
model/data set combination uniquely determines the LS estimates,
except for pathological cases, but some considerable computational
effort may nevertheless be required to arrive at the estimates. In
this book, the Gauss–Newton method is favored for reasons soon to be
stated, and all estimates will be determined using this procedure,
which is also known as the linearization method (Draper and Smith, 1981,
p. 462). Commonly used versions of this procedure, with modifications
designed to improve the speed of convergence, are those of Hartley
(1961) and Marquardt (1963). A more complete treatment of the
methodology is to be found in Bard (1974). We will not concern ourselves
here with a discussion of these methods since the focus of attention
will be on the *properties* of the resulting estimates. In addition,
as will become increasingly apparent as the chapters unfold, a cardinal
principle, to be emphasized repeatedly, is the desirability of adopting
models whose behavior closely approximates the behavior of linear
models. For such close-to-linear models, there will almost always be
a unique minimum and the speed of convergence of the unmodified
Gauss–Newton method to that minimum will usually be very rapid. This
follows from the fact that for a linear model the sum of squares surface
has a single minimum and the Gauss–Newton method will find that minimum
in a single iteration from any set of initial parameter estimates (see
Appendix 2.A). For a model which closely approximates linear behavior,
convergence will require more than a single iteration, but will
nevertheless be very rapid and not depend strongly on the initial
values. As the behavior of a model becomes more and more nonlinear,
convergence may not even occur. The response surface may have more
than one minimum, and convergence to the right minimum is never
guaranteed. Nevertheless, to examine the behavior of the LS estimator
using the techniques to be described in Secs. 2.4–2.8, the modeler must

be able to obtain the LS estimates for those badly behaved nonlinear models as well as close-to-linear ones. Rather than modify the Gauss–Newton method or seek an alternative method, it is better to direct attention toward obtaining good initial estimates of the parameters. This is the subject of Chapter 8, where procedures for obtaining good initial parameter estimates are presented for all models considered in Chapters 3–5. A description of the unmodified Gauss–Newton method is given in Appendix 2.A.

In addition to obtaining the LS estimates of the parameters, the modeler will wish to obtain an estimate of the asymptotic covariance matrix of the parameter estimates; this is needed for calculations to be described in Secs. 2.5 and 2.8. Most Gauss–Newton algorithms produce as a by-product an estimate of the large-sample covariance matrix, being the inverse of the "information matrix" evaluated at the converged values of the LS estimates. If the regression model is written as

$$Y_t = f(\underset{\sim}{X}_t, \underset{\sim}{\theta}) + \epsilon_t \qquad (2.1)$$

where $\underset{\sim}{\theta} = (\theta_1, \theta_2, \ldots, \theta_p)^T$ is the vector of parameters to be estimated, the input or regressor variables $\underset{\sim}{X}_t$ ($t = 1, 2, \ldots, n$) are, in general, k-dimensional vectors whose values are assumed known, and the errors ϵ_t ($t = 1, 2, \ldots, n$) are iidN with mean zero and unknown variance σ^2, then the elements of the information matrix are simply

$$\frac{1}{\sigma^2} \sum_t \frac{\partial f(\underset{\sim}{X}_t, \underset{\sim}{\theta})}{\partial \theta_i} \frac{\partial f(\underset{\sim}{X}_t, \underset{\sim}{\theta})}{\partial \theta_j} \qquad (2.2)$$

where θ_i and θ_j are elements of the parameter vector $\underset{\sim}{\theta}$ (i, j = 1, 2, …, p). The estimate of the information matrix is obtained at convergence by substituting the LS estimate $\hat{\underset{\sim}{\theta}}$ in place of $\underset{\sim}{\theta}$ and the estimated residual variance $\hat{\sigma}^2$ in place of $\overset{\sim}{\sigma}^2$, with $\hat{\sigma}^2$ obtained as RSS/(n − p), where RSS is the residual sum of squares, and n − p is the residual degrees of freedom.

Each diagonal element of the inverse of the matrix formed from the elements of (2.2) is a minimum variance bound for the corresponding parameter and is appropriate to a class of estimators obeying certain regularity conditions (Kendall and Stuart, 1967, Chap. 17). This is the minimum possible variance that was mentioned in Chapter 1 and which, in practice, provides the nonlinear modeler with one standard by which the performance of a model/data set combination may be judged.

2.3 Goodness of Fit of a Model

A question that has occupied the attention of modelers is that of how well some specified model fits the data. Extensive methodology has been developed for investigating whether a proposed model provides a good description of the data (e.g., see Daniel and Wood, 1971). The methods usually involve examination of the residuals, the $\hat{\varepsilon}_t$, these being the differences between the observed responses Y_t and the fitted or predicted responses $\hat{Y}_t = f(\underset{\sim}{X}_t, \hat{\theta})$. Daniel and Wood (1971, App. 3A) show cumulative distribution plots of generated random normal deviates with mean zero and unit variance. Sets of 16 are characterized by very obvious patterns. Although sets of 32 are visibly better behaved, they still exhibit patterns which could easily be assumed by the modeler to be departures from randomness even though the data are *known* to be random. Hence, in practice, a modeler is very apt to be misled, for small sample sizes, if a decision on goodness of fit is based solely on examination of the residuals (incidentally, the residuals are not independent, as there are only n – p degrees of freedom among them).

In view of the above, no attention will be paid in this book to the assessment of goodness of fit for models in combination with small data sets. The usual modeling situation is that a model is adopted because some theory and/or empirical evidence from the use of that model over many data sets indicates that that model is appropriate. There are circumstances, however, in which the use of a model is not particularly well founded, and several competing models may appear to fit the data equally well in practice. Here the choice between models must rest on other considerations; in general, the model that comes

closest to behaving as a linear model will be the preferred choice.
Of course, a model that does not fit the data will be of no interest
to the modeler, even if it may be shown to have various other desirable
properties.

2.4   The Curvature Measures of Nonlinearity of Bates and Watts

In Sec. 1.3 it was shown that the solution locus could be displayed
pictorially by plotting the observed data for a model/data set
combination in n-dimensional sample space. If $\hat{\underline{\theta}}$ represents the LS
estimator, the curvature of the solution locus in the vicinity of $\hat{\underline{\theta}}$
is a measure of the intrinsic nonlinearity, while the unequal spacing
and lack of parallelism of parameter lines projected onto the tangent
plane to the solution locus is a measure of parameter-effects
nonlinearity (following Bates and Watts, 1980).   These authors
quantified the measures by employing three-dimensional "acceleration
arrays." Their theoretical development is outlined in an appendix to
this chapter (Appendix 2.B).   For the intrinsic nonlinearity, the
acceleration array is of dimension $(n - p) \times p \times p$, and these $(n - p)p^2$
elements reduce to a single measure represented by the "maximum
intrinsic curvature," denoted IN in this book.   For the parameter-
effects nonlinearity, the corresponding acceleration array is of
dimension $p \times p \times p$, and these p^3 elements reduce to a single measure
represented by the "maximum parameter-effects curvature," denoted PE
in this book.   A FORTRAN program for carrying out these calculations
is presented in the Appendix to this book.

     The statistical significance of IN and PE may be assessed by
comparing these values with $1/(2\sqrt{F})$, where $F = F(p, n - p; \alpha)$ is obtained
from a table of the F-distribution corresponding to significance level
α.   The value $1/\sqrt{F}$ may be regarded as the radius of curvature of the
$100(1 - \alpha)\%$ confidence region, so the solution locus may be considered
to be sufficiently linear over an approximate 95% confidence region
if $IN < 1/(2\sqrt{F})$. Similarly, if $PE < 1/(2\sqrt{F})$, the projected parameter
lines of $\underline{\theta}$ may be regarded as being sufficiently parallel and uniformly
spaced.

The data set of Table 1.1, in combination with model (1.14), $E(Y) = X^\theta$, will serve to illustrate the method. Here $p = 1$, $n = 2$, and $n - p = 1$. To calculate the standard radius $\rho = \hat{\sigma}\sqrt{p}$ (see Appendix 2.B), one needs the residual variance $\hat{\sigma}^2$. This is $[(2.5 - 4.1517)^2 + (10 - 9.5468)^2]/1 = 2.9334$. Hence $\rho = 1.7127$. The scaled first and second partial derivative matrices are, respectively,

$$V_. = \frac{1}{1.7127} \begin{pmatrix} (\log 2)2^{2.0537} \\ (\log 3)3^{2.0537} \end{pmatrix} = \begin{pmatrix} 1.6802 \\ 6.1238 \end{pmatrix}$$

$$V_{..} = \frac{1}{1.7127} \begin{pmatrix} (\log 2)^2 2^{2.0537} \\ (\log 3)^2 3^{2.0537} \end{pmatrix} = \begin{pmatrix} 1.1646 \\ 6.7277 \end{pmatrix}$$

The matrix $V_.$ is decomposed to give

$$V_. = QR = Q\left(-\frac{\tilde{R}}{0}--\right)$$

$$= \begin{pmatrix} -.26460 & -.96438 \\ -.96438 & .26460 \end{pmatrix} \begin{pmatrix} -6.3500 \\ 0 \end{pmatrix}$$

The inverse of $\tilde{R}$ is $L = (\tilde{R})^{-1} = (-6.3500)^{-1} = -.15748$, and

$$U_{..} = L^T V_{..} L = -.15748 \begin{pmatrix} 1.1646 \\ 6.7277 \end{pmatrix} (-.15748)$$

$$= \begin{pmatrix} .028882 \\ .16685 \end{pmatrix}$$

Therefore, the combined acceleration array $A_{..}$ is

$$A_{..} = Q^T U_{..} = \begin{pmatrix} -.26460 & -.96438 \\ -.96438 & .26460 \end{pmatrix} \begin{pmatrix} .028882 \\ .16685 \end{pmatrix}$$

$$= \begin{pmatrix} -.1685 \\ \hline .0163 \end{pmatrix} \quad \begin{matrix} \leftarrow \text{ The first p rows relate to the} \\ \text{parameter-effects curvature} \\ \leftarrow \text{ The last } n - p \text{ rows relate to the intrinsic} \\ \text{curvature} \end{matrix}$$

For a single-element acceleration array, the maximum curvature is just the absolute value of the element. Hence the following values of the curvature measures are obtained:

IN = .0163

PE = .1685

From a table of the F-distribution, F(1, 1; .05) = 161.4, so $1/(2\sqrt{F})$ = .0394. Since IN < $1/(2\sqrt{F})$, the intrinsic nonlinearity of the solution locus is adequately small, but as PE > $1/(2\sqrt{F})$ the "uniform coordinates" assumption is rejected. This latter result is consistent with the appearance of the coordinate spacing in Fig. 1.2a, where values of θ having equal increments appear to be rather unequally spaced.

For model (1.15), a reparameterization of model (1.14) by means of the function ϕ = exp (θ), the following values of the curvature measures are obtained:

IN = .0163

PE = .0111

The measure of intrinsic nonlinearity is the same as before, as reparameterization does not alter the shape of the solution locus (see Exercise 1.3), but it is seen that the parameter-effects nonlinearity has been greatly reduced, and as PE < $1/(2\sqrt{F})$, the uniform coordinates assumption is now acceptable. This is consistent with the apparent spacing of the coordinates in Fig. 1.2b.

2.5 The Bias Calculation of M.J. Box

The formula derived by Box (1971) for calculating the bias in the ML (here also LS) estimates of the parameters in nonlinear regression models having a single response variate is

$$\text{Bias } (\hat{\underset{\sim}{\theta}}) = \frac{-\sigma^2}{2} \left(\sum_{u=1}^{n} \underset{\sim}{F}_u \underset{\sim}{F}_u^T \right)^{-1} \sum_{t=1}^{n} \underset{\sim}{F}_t \text{ tr} \left[\left(\sum_{u=1}^{n} \underset{\sim}{F}_u \underset{\sim}{F}_u^T \right)^{-1} \underset{\sim}{H}_t \right]$$

(2.3)

where $\underset{\sim}{F}_t$ (= $\underset{\sim}{F}_u$) is the p × 1 vector of first derivatives of $f(\underset{\sim}{X}_t, \underset{\sim}{\theta})$ and $\underset{\sim}{H}_t$ is the p × p matrix of second derivatives with respect to each of the elements of $\underset{\sim}{\theta}$, evaluated at $\underset{\sim}{X}_t$, where t = 1, 2, ..., n, and at the unknown true parameter values. In practice, $\hat{\theta}$ and $\hat{\sigma}^2$ are usually used in place of the unknown quantities. The bias given by the left-hand side of (2.3) is a $p \times 1$ vector representing the discrepancy between the estimates of the parameters and the true parameter values. Bates and Watts (1980) have shown that the bias in the LS estimates of the jth parameter (i.e., in the element $\hat{\theta}_j$ of $\hat{\theta}$) is simply a scaled version of the average diagonal element of a component of the parameter-effects acceleration array. Hence the bias calculation has been incorporated into the computer program for obtaining the curvature measures of Bates and Watts (see the Appendix of this book). Although the program may be used routinely for all calculations, a sample calculation is presented below for the data set of Table 1.1. Since $p = 1$, many of the matrix multiplications in (2.3) result in scalar quantities, simplifying the presentation. For model (1.14), $\hat{\theta} = 2.0537$ and the residual variance $\hat{\sigma}^2 = 2.9334$. In addition, the variance of the LS estimate $\hat{\theta}$ is Var $(\hat{\theta}) = .02480$, from the inverse of the information matrix (2.2).

Also, $F_t = (\log X_t)X_t^\theta$, so $F_1 = (\log 2)2^{2.0537} = 2.8777$ and $F_2 = (\log 3)3^{2.0537} = 10.4883$. Therefore, $\sum FF^T = 118.285$ and $(\sum FF^T)^{-1} = .008454$. Since $H_t = (\log X_t)^2 X_t^\theta$, $H_1 = 1.9947$ and $H_2 = 11.523$. Hence

$$\sum F_t \, tr \, (\sum FF^T)^{-1}H_t = [2.8777(1.9947) + 10.4883(11.523)](.008454)$$

$$= 1.0702$$

Therefore,

$$\text{Bias } (\hat{\theta}) = \frac{-2.9334(.008454)(1.0702)}{2} = -.01327$$

In many of the examples to be presented in Chapters 3-6, the *percentage bias*, the bias expressed as a percentage of the LS estimate, is a useful quantity as an absolute value in excess of 1% appears to be a good rule of thumb for indicating nonlinear behavior. In the example above, the percentage bias in $\hat{\theta}$ is

$$\% \text{ Bias } (\hat{\theta}) = \frac{-.01327(100)}{2.0537} = -.646\%$$

We have seen in the preceding section that model (1.14) behaves
nonlinearly, so the rule of thumb of 1% does not work for this simple
example. However, it works sufficiently often for it to be a valuable
guide to nonlinear behavior.

Since throughout this book we will be emphasizing the need to
search for better parameterizations, it will be useful to have a formula
which predicts the biases in the estimates of the parameters of the
new parameterization directly from knowledge of the biases in the
estimates of the parameters of the old parameterization. Hence, if

$$\phi = g(\underset{\sim}{\theta}) \tag{2.4}$$

so that the new parameter ϕ is a function of one or more of the elements
θ_j of $\underset{\sim}{\theta}$ (j = 1, 2, ..., p), then using derivations similar to those
employed by Box (1971), it may be shown that

$$\text{Bias } (\hat{\phi}) = \underset{\sim}{G}^T \text{ Bias } (\hat{\underset{\sim}{\theta}}) + \frac{1}{2} \text{tr } [\underset{\sim}{M} \text{ Cov } (\hat{\underset{\sim}{\theta}})] \tag{2.5}$$

where $\underset{\sim}{G}$ is the p × 1 vector of the first derivatives and $\underset{\sim}{M}$ is the p × p
matrix of second derivatives of $g(\theta)$ with respect to $\underset{\sim}{\theta}$, these
derivatives being evaluated at $\hat{\underset{\sim}{\theta}}$. Bias $(\hat{\underset{\sim}{\theta}})$ is the p × 1 vector of biases
from (2.3), and Cov $(\hat{\underset{\sim}{\theta}})$ is the p×p asymptotic covariance matrix of $\hat{\underset{\sim}{\theta}}$,
obtained from the inverse of the information matrix (2.2). The
left-hand side of (2.5) is a scalar giving the bias in $\hat{\phi}$.

In a similar fashion, the asymptotic variance of $\hat{\phi}$ may be obtained
from the asymptotic covariance matrix of $\hat{\underset{\sim}{\theta}}$ from the following general
formula:

$$\text{Var } (\hat{\phi}) = \text{tr } [(\underset{\sim\sim}{GG}^T) \text{ Cov } (\hat{\underset{\sim}{\theta}})] \tag{2.6}$$

When p = 1, $\underset{\sim}{\theta}$ may be represented by θ, and formula (2.5) reduces to

$$\text{Bias } (\hat{\phi}) = \text{Bias } (\hat{\theta}) \frac{\partial g(\theta)}{\partial \theta} + \frac{1}{2} \text{Var } (\hat{\theta}) \frac{\partial^2 g(\theta)}{\partial \theta^2} \tag{2.7}$$

and (2.6) reduces to

$$\text{Var } (\hat{\phi}) = \text{Var } (\hat{\theta}) \left[\frac{\partial g(\theta)}{\partial \theta} \right]^2 \qquad (2.8)$$

To avoid introducing new notation, it is understood that terms such as $\partial g(\theta)/\partial \theta$ and $\partial^2 g(\theta)/\partial \theta^2$ are to be evaluated at $\hat{\theta}$. Applying the foregoing formulas to the reparameterization (1.15) for the data set of Table 1.1, the following are obtained:

$$\phi = g(\theta) = \exp (\theta) = 7.7966$$

$$\frac{\partial g(\theta)}{\partial \theta} = \frac{\partial^2 g(\theta)}{\partial \theta^2} = \exp (\theta) = 7.7966$$

when evaluated at $\hat{\theta} = 2.0537$. Hence, from (2.7), the bias in the new parameter estimate is

$$\text{Bias } (\hat{\phi}) = -.01327(7.7966) + \frac{1}{2}(.02480)7.7966$$

$$= -.006789$$

Expressed as a percentage of $\hat{\phi}$, the bias is

$$\% \text{ Bias } (\hat{\phi}) = \frac{-.006789(100)}{7.7966} = -.087\%$$

Hence the bias in $\hat{\phi}$ is not only smaller than the bias in $\hat{\theta}$ in the absolute sense, it is also smaller in the relative sense. The question of how closely the distribution of $\hat{\phi}$ approaches a normal distribution will be explored in the next section.

From (2.8), the estimated asymptotic variance of $\hat{\phi}$ is

$$\text{Var } (\hat{\phi}) = (.02480)(7.7966)^2 = 1.5075$$

2.6 Simulation Studies

The curvature measures of nonlinearity of Bates and Watts (1980) described in Sec. 2.4 are objective measures of the intrinsic and parameter-effects nonlinearity in a model/data set combination. If the intrinsic nonlinearity is acceptably low, there may nevertheless be significant parameter-effects nonlinearity and the modeler may wish to seek an appropriate reparameterization. The parameter-effects measure offers no guide to a suitable reparameterization, and in a multiparameter situation does not identify which parameter or parameters may be mostly responsible for the nonlinear behavior. The bias calculation of Box (Sec. 2.5) can help identify which parameters have the greatest bias, while a simulation study may reveal the full extent of the non-normal behavior of the estimators and possibly suggest useful reparameterizations. In addition, it was pointed out in Sec. 1.4 that one of the purposes of this book is to investigate the nonlinear behavior of a model by examining the sampling properties of the LS estimator. The idea was advanced there that if the estimator has small bias, a distribution close to normal and a variance close to the minimum variance, it seems reasonable to speak of the estimator as being *close to linear* in its behavior. Simulation studies are probably the most direct and best way to enable the modeler to study the sampling properties of the LS estimator.

In a typical simulation study, it is assumed that both the form of the deterministic component of the model and the nature of the stochastic component ϵ are known. Data are generated using a set of predetermined values of the parameters, allowing only the values of ϵ to change randomly or pseudo-randomly from set to set. By this means, many sets of simulated data are produced, each of which provides a set of LS estimates of the parameters which may then be examined for their bias, variance, and distributional properties. The question of how many sets of data to simulate in any single simulation study must be answered. The author has found, as a result of considerable experience in simulating data for studying the sampling behavior of nonlinear regression models, that at least 500 and preferably 1000 pseudo-random data sets should be generated in each simulation study. Fewer samples may lead to misleading results, especially for statistics based on higher sample moments, such as skewness and kurtosis coefficients. Throughout this book, reported results of simulation studies will always be based on 1000 sets of simulated data.

Recall that the regression model is given by (2.1), where ϵ_t is a random variable that may be generated to be stochastically independent and identically normally distributed with zero mean and constant variance σ^2. For given $\underset{\sim}{X}_t$, the Y_t will also, of course, have variance σ^2. As the purpose of the simulation study is to examine the behavior of the LS estimators, the values used for $\underset{\sim}{\theta}$ and σ^2 to generate the Y_t will usually be $\hat{\underset{\sim}{\theta}}$ and $\hat{\sigma}^2$ obtained from the least-squares fit to the original data set, as generally they are the only available estimates for these unkown quantities. Hence the assumption is made, for the purposes of the simulation study, that the "true" values of $\underset{\sim}{\theta}$ and σ^2 are $\hat{\underset{\sim}{\theta}}$ and $\hat{\sigma}^2$, respectively.

Each set of simulated data is then fitted by least squares in the usual way, using the Gauss-Newton algorithm, a computer program for which appears in the Appendix to this book. As an initial estimate of $\underset{\sim}{\theta}$, it will usually suffice to take the assumed true value $\hat{\underset{\sim}{\theta}}$. Some models require more sophisticated techniques for obtaining good initial estimates, and this is treated in Chapter 8. If convergence to the LS estimates occurs successfully for each of the 1000 data sets in the simulation study, the user will have 1000 estimates for each parameter. Taking each parameter separately, one may calculate the first four sample moments of the set of estimates, thereby obtaining the sample mean m_1, the sample variance m_2, the skewness coefficient $g_1 = m_3/m_2^{3/2}$, and the kurtosis coefficient m_4/m_2^2, where m_3 and m_4 are the third and fourth sample moments about the mean, respectively. In addition, an approach to examining the multivariate behavior of the LS estimator may be taken by calculating bivariate sampling moments from the estimates corresponding to parameter pairs, but this involves a higher degree of computational effort. In this book, examination of the sampling properties of the LS estimator will be restricted to each parameter separately, in the hope that such questions as to whether the estimator exhibits normal behavior (which really means *multivariate* normal behavior) can be answered by an examination of the *marginal* distribution of each parameter separately.

As already stated, the difference between the sample mean and the assumed true value of the parameter is an estimate of the bias, and we shall see that under many circumstances the bias may usefully be

reported as a percentage of the true value of the parameter. The sample
variance may be compared using a chi-square test with the asymptotic
variance given by the inverse of (2.2) to test whether it significantly
exceeds the latter. This "excess" variance may usefully be reported
as the percentage by which the sample variance exceeds the asymptotic
variance. The coefficients of skewness and kurtosis may be compared
with their expected values under sampling from a normal distribution
(0 and 3 respectively) to test whether the observed set of parameter
estimates is behaving normally. For a large simulation study, g_1 is
approximately normally distributed with mean zero and standard
deviation $(6/N)^{1/2}$ (Snedecor and Cochran, 1980, p. 79), so for N = 1000,
approximate two-tailed 5% and 1% critical values for g_1 are ±.152 and
±.200, respectively. For the kurtosis coefficient, it is best to
subtract the value 3 expected for a normal distribution to obtain the
so-called coefficient of "excess kurtosis" $g_2 = m_4/m_2^2 - 3$. For large
N, g_2 is approximately normally distributed with mean zero and standard
deviation $(24/N)^{1/2}$ (Snedecor and Cochran, 1980, p. 80), so for N =
1000, approximate two-tailed 5% and 1% critical values for g_2 are ±.304
and ±.399, respectively.

We now illustrate a simulation study using the data of Table 1.1
involving model (1.11) having a single parameter,

$$Y_t = X_t^\theta + \varepsilon_t$$

where ε_t is assumed to be normally distributed with mean zero and
variance σ^2. The "true" values of θ and σ^2 are taken to be 2.05369
and 2.93336, respectively, these being the quantities obtained from
the original estimation (five-place decimals being retained for
greater accuracy in subsequent calculations). As n = 2, each set of
data consists of two pairs of observations, viz., (X_1, Y_1) and (X_2, Y_2), respectively, ε_1 and ε_2 being obtained with a computer program
for generating independent random normal numbers. Some routines
produce random variables that are distributed as N(0, 1), in which case
the variates obtained have to be multiplied by $(2.93336)^{1/2}$ to produce
ε_t with the desired variance. Hence $Y_1 = 2^{2.05369} + \varepsilon_1$ and $Y_2 = 3^{2.05369}$

$+ \varepsilon_2$ are the response variates corresponding to $X_1 = 2$ and $X_2 = 3$, respectively. This "new" set of data is then fitted by least squares to obtain a "new" estimate $\hat{\theta}$ of θ. One thousand such sets of data are generated, which, in turn, give rise to 1000 estimates of $\hat{\theta}$. These are presented as a histogram in Fig. 2.1a, where each $\hat{\theta}$ has been standardized by subtracting from it the sample mean and then dividing by the sample standard deviation. The first four sample moments of $\hat{\theta}$, and quantities derived from these, are given in Table 2.1. The bias is obtained as follows. Since the assumed true value used in the simulations is 2.05369, the bias is $2.03811 - 2.05369 = -.01558$. This may be tested for significance using the true asymptotic variance which is the inverse of (2.2), which, in this one-parameter case, is simply the reciprocal of a scalar quantity as follows:

$$\sigma^2 \left[\sum_{t=1}^{n} \left(\frac{\partial f(X_t, \theta)}{\partial \theta} \right)^2 \right]^{-1} = 2.93336\{ [(\log 2)2^{2.05369}]^2$$
$$+ [(\log 3)3^{2.05369}]^2 \}^{-1}$$

$$= \frac{2.93336}{118.2849} = .024799$$

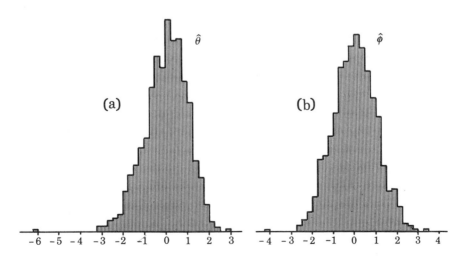

Fig. 2.1 Histograms of simulation results: (a) $\hat{\theta}$; (b) $\hat{\phi}$.

Table 2.1 Sample Moments and Derived Quantities for $\hat{\theta}$

Sample moments		Derived quantities	
Mean	2.03811	% Bias	$-.759^{**}$
Variance	.027088	% Excess variance	9.230^{*}
Third moment	$-.0026553$	Skewness	$-.596^{**}$
Fourth moment	.0030914	Excess kurtosis	1.213^{**}

Note: *P < .05; **P < .01.

The standard error of the sample mean from 1000 simulations is $(.024799/1000)^{1/2}$ = .004980. The bias $-.01558$ thus corresponds to an approximate standard normal deviate of $-.01558/.004980 = -3.129$, which is significant at the .01 level. The asymptotic variance may also be used to test whether the sample variance from the simulation study is significantly in excess of the minimum variance bound. Thus the statistic χ^2 = 999(.027088)/.024799 = 1091.2 is distributed as chi-square with 999 degrees of freedom (df), a transform of which may be closely approximated by the standard normal distribution, yielding

$$Z = \sqrt{2\chi^2} - \sqrt{2(df) - 1} = 46.716 - 44.688 = 2.028 \ (P < .05)$$

The result for the bias is presented in Table 2.1 as a percentage of the true parameter value, that is, $100(-.01558)/2.05369 = -.759\%$. Similarly, the result for the variance is presented as the percentage by which it is in excess of the assumed true value, that is, $100(.027088 - .024799)/.024799 = 9.230\%$. The skewness coefficient is $-.0026553/(.027088)^{3/2} = -.596$ and the coefficient of excess kurtosis is $.0030914/(.027088)^2 - 3 = 1.213$.

Having a set of results from a simulation study makes it a simple matter to examine the effect of reparameterization, since the LS estimate of a function of a parameter is simply that function of the LS estimate of the parameter. For example, to study the performance of parameter ϕ in the reparameterization given by model (1.15), one merely has to convert the 1000 estimates of $\hat{\theta}$ into 1000 estimates of

$\hat{\phi}$ via the function $\hat{\phi}$ = exp ($\hat{\theta}$). Results for $\hat{\phi}$, standardized by subtracting from it its sample mean and dividing by the sample standard deviation, are presented as a histogram in Fig. 2.1b. The first four sample moments of $\hat{\phi}$, and quantities derived from these, are given in Table 2.2.

It is clear from the coefficients of skewness and excess kurtosis in Table 2.2 that the distribution of $\hat{\phi}$ is close to normal. In addition, the observed bias and variance are not significantly different from their true values.

The closeness of approach to a normal distribution of the set of 1000 parameter estimates can also be visually assessed by examining the histograms in Fig. 2.1a and b for $\hat{\theta}$ and $\hat{\phi}$, respectively, where the estimates have been standardized to have zero mean and unit variance with class intervals of .25. For $\hat{\phi}$, no obvious visible departure is apparent from that expected for a normally distributed random variable. The histogram for $\hat{\theta}$, however, quite clearly illustrates the negative skewness in the LS estimator of θ in this parameterization.

The simulation results also make it possible to test how close the observed bias is to the calculated bias using the formula of Box given by (2.3). Results are presented in Table 2.3 with percentage bias given in parentheses.

It will be seen from the examples given in later chapters (Chapters 3-6) that Box's calculation is very successful in predicting bias to the correct order of magnitude, especially when the bias is large.

Table 2.2 Sample Moments and Derived Quantities for $\hat{\phi}$

Sample moments		Derived quantities	
Mean	7.7776	% Bias	-.245[ns]
Variance	1.5201	% Excess variance	.838[ns]
Third moment	-.045978	Skewness	-.025[ns]
Fourth moment	6.8607	Excess kurtosis	-.031[ns]

Note: [ns]not significant, P > .05.

Table 2.3 Comparison of Bias Calculated from Box's Formula (2.3) with Bias from Simulation

	Parameter	
	θ	ϕ
Calculated bias	$-.01327$ ($-.646\%$)	$-.00679$ ($-.087\%$)
Simulation bias	$-.01558$ ($-.759\%$)	$-.01907$ ($-.245\%$)

2.7 Confidence Regions for the Parameters

Several methods for determining confidence regions for the parameters of nonlinear regression models have been proposed. Reviews of these are given by Gallant (1976) and Shaw and Griffiths (1979). One of the methods, based on the asymptotic properties of the likelihood ratio, provides contours whose boundaries coincide with likelihood contours, but in principle there may be a large discrepancy between the actual and nominal confidence levels associated with those contours. Good agreement has been found between the likelihood ratio method and an exact method (Gallant, 1976; Shaw and Griffiths, 1979); as LS estimators are also ML estimators when the stochastic term is iidN, this agreement helps justify using the likelihood ratio method, which is based on the same principle of maximizing likelihood. This method defines the boundaries of a $100(1 - \alpha)\%$ confidence region for $\underset{\sim}{\theta}$ as those values of $\underset{\sim}{\theta}$ for which the following equality holds:

$$S(\underset{\sim}{\theta}) = S(\hat{\underset{\sim}{\theta}}) \left[1 + \frac{p}{n - p} \, F(p, \ n - p; \ \alpha) \right] \qquad (2.9)$$

where $S(\hat{\underset{\sim}{\theta}})$ is the minimum residual sum of squares occurring, of course, at $\hat{\underset{\sim}{\theta}}$ and $S(\underset{\sim}{\theta})$ is the residual sum of squares occurring at any other value of $\underset{\sim}{\theta}$. The value $F = F(p, \ n - p; \ \alpha)$ is obtained from a table of the F-distribution corresponding to significance level α and numerator and denominator degrees of freedom p and n − p, respectively. As all terms on the right-hand side of (2.9) are known, $S(\underset{\sim}{\theta})$ is easily calculated for a given α. The $100(1 - \alpha)\%$ confidence region for $\underset{\sim}{\theta}$ is then the locus of values of $\underset{\sim}{\theta}$ which result in that value of $S(\underset{\sim}{\theta})$. Sophisticated techniques involving interpolation routines may be needed to compute the regions.

The use of (2.9) may be illustrated using the data set in Table 1.1. For model (1.14), p = 1 and the confidence region in this case becomes simply a confidence *interval*. For α = .05, F(p, n - p; α) = 161.4. As $S(\hat{\theta})$ = 2.93336, $S(\theta)$ = 476.38. Thus a 95% confidence interval for θ is obtained by determining the values of θ for which the following equality holds:

$$476.38 = (2.5 - 2^{\theta})^2 + (10.0 - 3^{\theta})^2 \qquad (2.10)$$

Numerical solution of this shows that the upper limit is 3.123, to three decimal places. There is no lower limit, for as $\theta \to -\infty$, the right-hand side of (2.10) is never larger than $(2.5)^2 + (10.0)^2 = 106.25$. Such "unbounded" confidence regions are not unusual in nonlinear regression models. Results are presented in Table 2.4 both for θ and for ϕ in the reparameterization given by model (1.15). The values of ϕ_L and ϕ_U may be simply determined from θ_L and θ_U, respectively, using $\phi_i = \exp(\theta_i)$.

Table 2.4   Confidence Regions for Parameters $\hat{\theta}$ and $\hat{\phi}$ in (1.14) and (1.15), Respectively

			$\hat{\theta}$ = 2.0537		$\hat{\phi}$ = 7.7966	
Significance level	Corresponding confidence interval (%)	$S(\theta)$ [= $S(\phi)$]	θ_L (lower limit)	θ_U (upper limit)	ϕ_L (lower limit)	ϕ_U (upper limit)
.05	95	476.4	–	3.123	–	22.726
.10	90	120.0	–	2.733	–	15.380
.25	75	20.03	1.559	2.374	4.756	10.745
.50	50	5.867	1.880	2.200	6.551	9.027

The asymmetry of θ_L and θ_U with respect to $\hat{\theta}$ is readily observed in the result for the 75% confidence interval. In contrast, the results for ϕ_L and ϕ_U are much closer to being symmetric about $\hat{\phi}$, reflecting the fact that model (1.15) is more "linear" in its behavior than model (1.14), since for a linear model, the upper and lower limits are exactly symmetric about the LS estimate.

For $p = 2$, a two-parameter model, the extent to which a contour calculated using (2.9) departs from an ellipse (the condition for a linear model) provides a visual picture of the degree of nonlinearity in the model. These contours may be determined by evaluating $S(\theta)$ at the points of a rectangular grid corresponding to the components (θ_1, θ_2) of $\underset{\sim}{\theta}$ spread in both directions around $\hat{\underset{\sim}{\theta}}$. A sophisticated contouring routine may be needed to interpolate between points of the grid to produce smooth contours of a given $S(\theta)$ value. For $p = 3$, a three-dimensional grid of points will be needed if one desires to display the contours in three dimensions. The closeness of approach of the contours to an ellipsoid is a measure of the extent to which a nonlinear model approaches linear behavior. Alternatively, the results may be displayed as three two-dimensional diagrams, in which two of the components $(\theta_1, \theta_2, \theta_3)$ of $\underset{\sim}{\theta}$ are allowed to vary in turn, while the third component is held constant at the LS estimate. For $p > 3$, there is little alternative other than to use the latter method of presenting the results as $p(p - 1)/2$ two-dimensional diagrams. The closer the model is to being linear, the closer will the contours of the two-dimensional diagrams approach elliptical shape. However, this procedure has its disadvantages because even a badly asymmetric "banana-shaped" region in multidimensional space, when sectioned through the LS estimate, will produce two-dimensional contours that tend to appear far more symmetric than in the space of full dimension. A further disadvantage is that there is as yet no readily available criterion or statistical test to assess the significance of the degree of asymmetry in a confidence contour. Coupled with the fact that confidence regions are very expensive of computation time, this renders their calculation to be of less practical value for assessing nonlinearity than the techniques described in Secs. 2.4-2.6, especially for the three-parameter and four-parameter models which make up so many of the examples considered in Chapters 3-6.

2.8 t-Values for the Parameter Estimates

A criterion that might be deemed potentially useful for examining the behavior of a model in estimation is the value of Student's t associated with the parameter estimate, defined as

$$t = \frac{\hat{\theta}}{Var^{1/2}(\hat{\theta})} \tag{2.11}$$

That is, the t-value is the ratio of the parameter estimate to its standard error, the latter being estimated by the square root of the asymptotic variance of the estimate. Computer software packages based on the Gauss-Newton algorithm can routinely produce estimates of the asymptotic variances and the t-values, as is the case with the program in the Appendix to this book. A high t-value associated with a parameter estimate tends to indicate that the estimate is well determined in the model; conversely, a low t-value tends to indicate that the estimate is poorly determined, although sometimes in multiparameter models, a t-value may be low because of high correlation of the parameter with other parameters in the model. The t-values may be tested by reference to a Student's t-distribution with n-p degrees of freedom, the degrees of freedom associated with the residual variance $\hat{\sigma}^2$.

Use of the t-values will be illustrated by employing the data set of Table 1.1. Consider first model (1.14), for which the following parameter estimate and estimate of asymptotic variance were obtained:

$$\hat{\theta} = 2.0537$$
$$Var\ (\hat{\theta}) = .02480$$

The t-value, obtained by applying (2.11), is

$$t = \frac{\hat{\theta}}{Var^{1/2}(\hat{\theta})} = 13.04 \quad (1\ df,\ P < .05)$$

The t-value is relatively high and statistically significant at α = .05.

Consider now model (1.15), being a reparameterization of model (1.14) via the function ϕ = exp (θ), for which the following parameter estimate and estimate of asymptotic variance were obtained:

$$\hat{\phi} = 7.7966$$
$$\text{Var } (\hat{\phi}) = 1.5075$$

The t-value, obtained using (2.11), is

$$t = \frac{\hat{\phi}}{\text{Var}^{1/2}(\hat{\phi})} = 6.35 \quad (1 \text{ df, } P < .05)$$

Thus, the t-value for the reparameterized parameter ϕ in model (1.15) is smaller and less significant statistically than the t-value for θ in the original parameterization of model (1.14) although of the two models, the reparameterized model was found to be the more close to linear in its estimation behavior by the techniques of Secs. 2.4-2.7. This illustrates that although the t-values may be useful in indicating whether or not a parameter may be well determined or badly determined in a model, it should be used with caution when trying to assess a model's behavior in estimation. Although low t-values will usually indicate that a parameter is poorly determined, meaning that confidence intervals or confidence regions involving that parameter may include zero, a high t-value does not necessarily indicate that the parameter estimate will exhibit desirable statistical properties such as a close approach to linear behavior in estimation. It will be seen in Sec. 4.5, however, that the magnitudes of the t-values *can* be useful indicators of nonlinear behavior.

2.9 Asymmetry Measure of Bias

We now propose a new measure of bias and nonlinearity. The fundamental idea and resulting statistic is due to Robert K. Lowry. At the time of writing, the properties of the statistic are still under investigation, but preliminary results suggest that the statistic is a useful tool in indicating nonlinear behavior of the LS estimator.

Consider the regression model

$$Y = f(X, \underset{\sim}{\theta}) + \varepsilon \qquad (2.12)$$

where $f(X,\theta)$ may be linear or nonlinear in the parameter vector $\underset{\sim}{\theta}$ $(= \theta_1, \theta_2, \ldots, \theta_p)$. Assuming that $\underset{\sim}{\theta}$ is known, one may generate a series of iidN errors ε_t (t = 1,2, $\ldots$, n) having mean zero and variance σ^2 corresponding to X_t (t = 1,2, $\ldots$, n), resulting in a set of random variables denoted Y_t^+, that is

$$Y_t^+ = f(X_t, \underset{\sim}{\theta}) + \varepsilon_t \qquad (2.13)$$

The LS estimate of $\underset{\sim}{\theta}$ corresponding to this generated model/data set combination is denoted $\hat{\underset{\sim}{\theta}}^+$. Now return to the regression model (2.12) and consider the random variables obtained by *subtracting* the generated ε_t from $f(X_t,\theta)$ instead of *adding* the generated ε_t to $f(X_t,\theta)$. That is, we have the following set of dependent variables, denoted Y_t^-:

$$Y_t^- = f(X_t, \underset{\sim}{\theta}) - \varepsilon_t \qquad (2.14)$$

The LS estimates of $\underset{\sim}{\theta}$ corresponding to this new data set is denoted $\hat{\underset{\sim}{\theta}}^-$.

For a linear model, it is true that for each parameter θ_i (i = 1,2, $\ldots$, p),

$$(\hat{\theta}_i^+ - \theta_i) = -(\hat{\theta}_i^- - \theta_i) \qquad (2.15)$$

Viewed geometrically, the estimates $\hat{\theta}_i^+$ and $\hat{\theta}_i^-$ are symmetric about the true parameter value θ_i. For a nonlinear model, however, (2.15) is in general not true. It follows then that the statistic

$$\psi_i = \psi(\hat{\theta}_i^+ \, \hat{\theta}_i^-) = \frac{(\hat{\theta}_i^+ - \theta_i) + (\hat{\theta}_i^- - \theta_i)}{2} \qquad (2.16)$$

for i = 1, 2, $\ldots$, p, can be used as a measure of nonlinear behavior of the LS estimator of θ_i. Since the distribution of $\hat{\theta}_i^+$ is identical with the distribution of $\hat{\theta}_i^-$, each being an alternative LS estimator of θ_i, denoted $\hat{\theta}_i$, it follows that

$$E(\psi_i) = E(\hat{\theta}_i^+ - \theta_i) = E(\hat{\theta}_i^- - \theta_i) = \text{Bias} \ (\hat{\theta}_i)$$

By substituting (2.15) into (2.16), it is seen that for a linear model, the expectation of ψ_i, which is the bias of $\hat{\theta}_i$, is zero. Similarly, the variance of ψ_i is zero for a linear model. For a nonlinear model, the variance of ψ_i is

$$\text{Var} \ (\psi_i) = \frac{1}{4} \ \text{Var} \ (\hat{\theta}_i^+) + \frac{1}{4} \ \text{Var} \ (\hat{\theta}_i^-) + \frac{1}{2} \ \text{Cov} \ (\hat{\theta}_i^+ \ \hat{\theta}_i^-) \qquad (2.17)$$

The correlation coefficient between $\hat{\theta}_i^+$ and $\hat{\theta}_i^-$ can be defined as

$$\text{Corr} \ (\hat{\theta}_i^+ \ \hat{\theta}_i^-) = \frac{\text{Cov} \ (\hat{\theta}_i^+ \ \hat{\theta}_i^-)}{\sqrt{\text{Var} \ (\hat{\theta}_i^+) \ \text{Var} \ (\hat{\theta}_i^-)}} \qquad (2.18)$$

For linear models, because Var (ψ_i) is zero and because Var $(\hat{\theta}_i^+)$ = Var $(\hat{\theta}_i^-)$ = Var $(\hat{\theta}_i)$, it follows that Corr $(\hat{\theta}_i^+ \hat{\theta}_i^-)$ = -1. For nonlinear models, the correlation will vary between -1 and 1, with close-to-linear models producing values close to -1.

We now illustrate the use of Eqs. (2.16)-(2.18) on the data set of Table 1.1 in combination with models (1.14) and (1.15). One thousand sets of simulated data are generated for model (1.14) as described in Sec. 2.6, and for each data set the LS estimate of θ, denoted $\hat{\theta}^+$, is obtained. For each set of data, the signs of ε_t (t = 1,2) are then reversed, and new values of the dependent variable are thus generated. The LS estimate of θ for each of these modified data sets is denoted $\hat{\theta}^-$. Using (2.16), ψ is calculated for each data set. The mean value $\bar{\psi}$ of ψ over the 1000 data sets is an estimate of the bias. In addition, the sample variance of ψ is obtained, as well as the sample variances of $\hat{\theta}^+$ and $\hat{\theta}^-$, using the 1000 sets of values for each of these estimators. Substituting into (2.17), Cov $(\hat{\theta}^+ \ \hat{\theta}^-)$ is determined and used to calculate Corr $(\hat{\theta}^+ \ \hat{\theta}^-)$ employing (2.18). From the 1000 sets of estimates of $\hat{\theta}^+$ and $\hat{\theta}^-$, values of $\hat{\phi}^+$ and $\hat{\phi}^-$ in model (1.15) are calculated using exp $(\hat{\theta}^+)$ and exp $(\hat{\theta}^-)$, respectively. Then, Eqs. (2.16)-(2.18) are applied to $\hat{\phi}^+$ and $\hat{\phi}^-$. Results are presented in Table 2.5.

Table 2.5 Results on the Asymmetry Measure Using Simulated Data for Models (1.14) and (1.15)

Parameter	θ in (1.14)	ϕ in (1.15)
Bias (2.16)	-.01452	-.007292
Correlation coefficient (2.18)	-.9661	-.9994

The bias of $\hat{\theta}$ in (1.14) of -.01452 using the asymmetry measure compares well with the value -.01327 obtained using the calculation of Box (Sec. 2.5); similarly, the bias of $\hat{\phi}$ in (1.15) of -.007292 using the asymmetry measure compares well with the value -.006789 from Box's formula. Since the correlation coefficient -.9994 for $\hat{\phi}$ using (2.18) is very close to unity in absolute value, $\hat{\phi}$ is close to linear in behavior, in agreement with the results of Secs. 2.4-2.7. The correlation coefficient -.9661 for $\hat{\theta}$ using (2.18) is lower, indicating that $\hat{\theta}$ is less close to linear in behavior than $\hat{\phi}$.

The new asymmetry measure appears to be a powerful tool in helping to elucidate which parameters have estimators that behave nonlinearly. It does not suffer from the same defects as the "percentage bias", discussed in Sec. 9.5. That is, the correlation coefficient using (2.18) is highly correlated with the nonlinear behavior of the estimator, even when the true value of the parameter is close to zero or when the parameter represents a constant term in the model. Further study of this measure is being undertaken by Dr. Richard Morton, CSIRO Division of Mathematics and Statistics, Canberra, who has obtained an expression for its variance using a Taylor series expansion. Further analytical results may be expected, although the resulting expressions are cumbersome. A simulation study, as used above, is an alternative approach for obtaining numerical results.

2.10 Some Remarks on Reparameterizations

Consider different parameterizations of a model, by which it is meant that the parameters of the new parameterization (or "reparameterization") are related to the parameters of the old parameterization by an expression of the form (2.4), an expression which involves *parameters only* and not the regressor variables $X_{\sim t}$. Bates and Watts (1980) refer to various parameterizations of the same model as "model functions" and we will also use this terminology. Two model functions of the same model studied extensively in this chapter for illustrative purposes are (1.15) and (1.14), the latter of course being the same as (1.11) but written differently. The methods described in Secs. 2.4-2.7 show that (1.15) behaves more like a linear model than (1.14), and is thereby the more desirable model function for statistical purposes. What about the acceptability of (1.15) to the modeler? It could be argued that if parameter θ in (1.14) had some fundamental significance (biological, physical, engineering, etc.), then this significance might be lost by the reparameterization $\phi = \exp(\theta)$. On the other hand, there are also many examples in pure and applied science in which transformations of the foregoing type are in wide use and are readily understood by, and highly acceptable to, the user. Two such examples are the decibel (dB) used in physics and engineering to quantify the loudness of sound, and pH, used in chemistry to quantify the acidity or alkalinity of solutions. The former is proportional to the logarithm of the sound pressure wave, whereas the latter is the logarithm of the reciprocal of the hydrogen ion concentration. Neither of these simple monotonic transformations presents any problems of understanding to the user, despite the fact that with pH there is the added complication that pH *decreases* as acidity *increases*, as a result of the reciprocal of concentration being used. Therefore, there is little reason to anticipate that reparameterization would encounter user resistance or any barriers to user understanding, provided that the relationship of the new parameters to the old ones generally satisfy a few simple criteria, such as monotonicity and single-valuedness.

The search for suitable reparameterizations, to produce model functions whose behavior in estimation is close to linear, plays a central role throughout this book. It would be desirable to have a set of simple rules to help the user in the search for reparameterizations. It is unfortunate that the curvature measure of parameter-effects nonlinearity does not identify the nonlinear-behaving parameters nor does it suggest suitable reparameterizations; hence the user must rely on a simulation study. From the histograms of the results for each parameter separately, an appropriate reparameterization may be deduced. For example, a histogram with a long right-hand tail characteristic of a lognormal distribution suggests replacement of the parameter in the model function by the exponential of the parameter. Conversely, a histogram with a long left-hand tail suggests replacement of the parameter by the logarithm of the parameter. Aside from these rather obvious procedures no general guidelines can be given. Fortunately, expression (2.5) is available to help the user evaluate the possible effect of a proposed reparameterization with a minimum of calculation.

Appendix 2.A: The Gauss-Newton Method

For a more detailed discussion of this method and its relationship to Newton's method (also called the Newton-Raphson method), the reader is referred to Kennedy and Gentle (1980, Chap. 10).

The nonlinear regression model is given by (2.1) as

$$Y_t = f(\underset{\sim}{X}_t, \underset{\sim}{\theta}) + \varepsilon_t$$

so that the sum of squares to be minimized is

$$S(\underset{\sim}{\theta}) = \sum_t [Y_t - f(\underset{\sim}{X}_t, \underset{\sim}{\theta})]^2$$

To simplify the notation, we write $f_t(\underset{\sim}{\theta})$ in place of $f(\underset{\sim}{X}_t, \underset{\sim}{\theta})$ and S in place of $S(\underset{\sim}{\theta})$. The n × p "Jacobian matrix" is

$$J(\theta^*) = \begin{bmatrix} \dfrac{\partial f_1(\theta)}{\partial \theta_1} & \dfrac{\partial f_1(\theta)}{\partial \theta_2} & \cdots & \dfrac{\partial f_1(\theta)}{\partial \theta_p} \\ & \bullet & & \bullet \\ & \bullet & & \bullet \\ & \bullet & & \bullet \\ \dfrac{\partial f_n(\theta)}{\partial \theta_1} & \dfrac{\partial f_n(\theta)}{\partial \theta_2} & \cdots & \dfrac{\partial f_n(\theta)}{\partial \theta_p} \end{bmatrix}_{\theta=\theta^*}$$

We may expand $f_t(\theta)$ in a Taylor series about θ_i (the subscript i indicating the ith iteration), and retain only the first two terms:

$$\underset{\sim}{f}(\theta) \cong \underset{\sim}{f}(\theta_i) + J(\theta_i)(\theta - \theta_i)$$

where $\underset{\sim}{f}(\theta) = [f_1(\theta), f_2(\theta), \ldots, f_n(\theta)]^T$

Writing $\underset{\sim}{Y} = [Y_1, Y_2, \ldots, Y_n]^T$, it follows that

$$[\underset{\sim}{Y} - \underset{\sim}{f}(\theta)]^T[\underset{\sim}{Y} - \underset{\sim}{f}(\theta)] \cong [\underset{\sim}{Y} - \underset{\sim}{f}(\theta_i)]^T[\underset{\sim}{Y} - \underset{\sim}{f}(\theta_i)]$$

$$- 2[\underset{\sim}{Y} - \underset{\sim}{f}(\theta_i)]^T J(\theta_i)(\theta - \theta_i)$$

$$+ (\theta - \theta_i)^T J^T(\theta_i) J(\theta_i)(\theta - \theta_i)$$

The gradient vector $g(\theta) = [\partial S/\partial \theta_1, \partial S/\partial \theta_2, \ldots, \partial S/\partial \theta_p]^T$ is therefore

$$g(\theta) = -2J^T(\theta_i)[\underset{\sim}{Y} - \underset{\sim}{f}(\theta_i)] + 2J^T(\theta_i)J(\theta_i)(\theta - \theta_i)$$

Equating this expression to zero, and rearranging, yields

$$\theta_{i+1} = \theta_i + [J^T(\theta_i)J(\theta_i)]^{-1}J^T(\theta_i)[\underset{\sim}{Y} - \underset{\sim}{f}(\theta_i)]$$

Starting with initial values for θ at i = 1, the process continues until convergence, which occurs when $|\theta_{i+1} - \theta_i|$ is smaller than some preselected small quantity. At convergence, the estimate of the asymptotic covariance matrix is

$$\text{Cov } (\hat{\underset{\sim}{\theta}}) = \hat{\sigma}^2 [J^T(\hat{\underset{\sim}{\theta}}) J(\hat{\underset{\sim}{\theta}})]^{-1}$$

We now consider the use of the Gauss–Newton method on a *linear* regression model. Such a model may be written in general as

$$\underset{\sim}{Y} = \underset{\sim}{X}\underset{\sim}{\theta} + \underset{\sim}{\epsilon}$$

where $\underset{\sim}{X}$ is the n × p matrix of regressor variables, whose first column may be a unit vector if the model function includes a constant term. For the model function above, the Jacobian matrix is

$$J(\underset{\sim}{\theta}) = \underset{\sim}{X}$$

Starting with any arbitrary set of initial parameter estimates $\underset{\sim}{\theta}_0$, the updated vector of estimates $\underset{\sim}{\theta}_1$ is

$$\begin{aligned}
\underset{\sim}{\theta}_1 &= \underset{\sim}{\theta}_0 + (\underset{\sim}{X}^T\underset{\sim}{X})^{-1}\underset{\sim}{X}^T[\underset{\sim}{Y} - \underset{\sim}{X}\underset{\sim}{\theta}_0] \\
&= \underset{\sim}{\theta}_0 + (\underset{\sim}{X}^T\underset{\sim}{X})^{-1}\underset{\sim}{X}^T\underset{\sim}{Y} - (\underset{\sim}{X}^T\underset{\sim}{X})^{-1}(\underset{\sim}{X}^T\underset{\sim}{X})\underset{\sim}{\theta}_0 \\
&= \underset{\sim}{\theta}_0 + (\underset{\sim}{X}^T\underset{\sim}{X})^{-1}\underset{\sim}{X}^T\underset{\sim}{Y} - I\underset{\sim}{\theta}_0 \\
&= \underset{\sim}{\theta}_0 + (\underset{\sim}{X}^T\underset{\sim}{X})^{-1}\underset{\sim}{X}^T\underset{\sim}{Y} - \underset{\sim}{\theta}_0 \\
&= (\underset{\sim}{X}^T\underset{\sim}{X})^{-1}\underset{\sim}{X}^T\underset{\sim}{Y}
\end{aligned}$$

The right-hand side, usually denoted by $\underset{\sim}{b}$ (Draper and Smith, 1981), is the LS estimator $\hat{\underset{\sim}{\theta}}$ for a linear regression model. It is obvious that the estimates remain unchanged if the iterative process is continued. Hence the Gauss–Newton method, for a linear model, converges to the LS estimators in a single iteration from any starting vector $\underset{\sim}{\theta}_0$. The estimated covariance matrix is

$$\text{Cov } (\hat{\underset{\sim}{\theta}}) = \hat{\sigma}^2 (\underset{\sim}{X}^T\underset{\sim}{X})^{-1}$$

which, unlike that for a nonlinear model, does not have an asymptotic character.

Appendix 2.B: The Curvature Measures of Nonlinearity

For a more detailed description of the methodology, readers are referred to the expository paper of Bates and Watts (1980). Only a brief outline is presented here.

For the regression model function with t sets of observations, t = 1, 2, ..., n, written as

$$Y_t = f(X_t, \theta) + \varepsilon_t$$

where $\theta = (\theta_1, \theta_2, \ldots, \theta_p)^T$ is the vector of parameters to be estimated, the conditional expected response is

$$\eta(\theta) = f(X_t, \theta)$$

The least-squares procedure minimizes

$$S(\theta) = \sum_t [Y_t - \eta_t(\theta)]^2$$

which may be written as

$$S(\theta) = \| Y - \eta_t(\theta) \|^2$$

where $\eta(\theta) = [\eta_1(\theta), \eta_2(\theta), \ldots, \eta_n(\theta)]^T$

and double vertical bars indicate the length of a vector.

Algorithms for computing the least-squares estimates $\hat{\theta}$ such as the Gauss-Newton method (Appendix 2.A) are based on a local linear approximation to the model function. In the vicinity of a fixed parameter value θ_0 the model function is approximated by

$$\eta(\theta) \cong \eta(\theta_0) + \sum_{i=1}^{p} (\theta_i - \theta_{i,0}) v_i$$

where $v_i = [v_i(x_1), v_i(x_2), \ldots, v_i(x_n)]^T$

with $v_i(x) = \dfrac{\partial \eta(\theta)}{\partial \theta_i}$ evaluated at θ_0

Thus, one can construct the $n \times p$ matrix V, consisting of first partial derivatives of the model function with respect to the parameters, evaluated at $\underset{\sim}{\theta}_0$, the ith column being $\underset{\sim}{v}_i$.

Similarly, one can construct the $n \times p \times p$ matrix $V_{..}$ of second partial derivatives with respect to the parameters. The elements of this matrix are

$$v_{tij} = \frac{\partial^2 n_t(\underset{\sim}{\theta})}{\partial \theta_i \, \partial \theta_j}$$

evaluated at $\underset{\sim}{\theta}_0$.

An arbitrary straight line in the parameter space through $\underset{\sim}{\theta}_0$ may be expressed using a geometric parameter b by

$$\underset{\sim}{\theta}(b) = \underset{\sim}{\theta}_0 + b\underset{\sim}{h}$$

where $\underset{\sim}{h} = (h_1, h_2, \ldots, h_p)^T$ is any nonzero vector. This line generates a curve or "lifted line" on the solution locus given by

$$\underset{\sim}{n}_h(b) = \underset{\sim}{n}(\underset{\sim}{\theta}_0 + b\underset{\sim}{h})$$

The tangent to this curve at $b = 0$ may be shown to be

$$\dot{\underset{\sim}{n}}_h = V_{.}\underset{\sim}{h}$$

and the set of all such linear combinations is referred to as the *tangent plane* at $\underset{\sim}{n}(\underset{\sim}{\theta}_0)$. The acceleration of the lifted line $\underset{\sim}{n}_h$ may be shown to be

$$\ddot{\underset{\sim}{n}}_h = \underset{\sim}{h}^T V_{..} \underset{\sim}{h}$$

so that this $n \times 1$ vector is just a linear combination of the $p \times p$ matrix of second partial derivatives $\underset{\sim}{v}_{ij}$ for each of the n observations. Each element of $\ddot{\underset{\sim}{n}}_h$ has the form $\underset{\sim}{h}^T V_t \underset{\sim}{h}$, where V_t is the tth face of the $V_{..}$ array.

The acceleration vector $\ddot{\underset{\sim}{n}}_h$ can be written as three components, the first component $\ddot{\underset{\sim}{n}}_h^{IN}$ determining the change in direction of the instantaneous velocity vector $\dot{\underset{\sim}{n}}_h$ normal to the tangent plane, whereas the second and third components, which can be added together to give $\ddot{\underset{\sim}{\eta}}_h^{PE}$, determine respectively the change in direction of $\dot{\underset{\sim}{\eta}}_h$ parallel to the tangent plane and the change in speed of the moving point. The acceleration components may be converted into curvatures to give the *intrinsic* curvature

$$
K_{\underset{\sim}{h}}^{IN} = \frac{\| \ddot{\underset{\sim}{\eta}}_h^{IN} \|}{\| \dot{\underset{\sim}{\eta}}_h \|^2}
$$

and the *parameter-effects* curvature

$$
K_{\underset{\sim}{h}}^{PE} = \frac{\| \ddot{\underset{\sim}{\eta}}_h^{PE} \|}{\| \dot{\underset{\sim}{\eta}}_h \|^2}
$$

Only the latter depends on the particular parameterization chosen. $K_{\underset{\sim}{h}}^{IN}$ and $K_{\underset{\sim}{h}}^{PE}$ may be converted to response-invariant standardized relative curvatures $\gamma_{\underset{\sim}{h}}^{IN}$ and $\gamma_{\underset{\sim}{h}}^{PE}$, respectively, by multiplying by the *standard radius*

$$
\rho = \hat{\sigma}\sqrt{p}
$$

where $\hat{\sigma}$ is the square root of the estimated residual variance $\hat{\sigma}^2$. The curvatures may be calculated directly using scaled data, where the elements of $V_{\boldsymbol{.}}$ and $V_{\boldsymbol{..}}$ have been divided by ρ. In what follows, it is assumed that this scaling has been done.

The curvature calculations can be simplified by first making a coordinate transformation which rotates the coordinates of the sample space so that the first p coordinate vectors are parallel to the tangent plane and the last $n - p$ are orthogonal to it. This is done by premultiplying all the vectors in sample space by an orthogonal matrix Q^T, where Q is part of the QR decomposition of $V_{\boldsymbol{.}}$, that is,

$$
\underset{n \times p}{V_{\boldsymbol{.}}} = \underset{n \times n}{Q} \quad \underset{n \times p}{R} = Q\left(\frac{\tilde{R}}{0}\right) \begin{array}{l} \leftarrow \ p \times p \\ \leftarrow \ (n - p) \times p \end{array}
$$

where $\tilde{R}$ is upper triangular and 0 the zero matrix.

The coordinates in the parameter space are also transformed from $\underset{\sim}{\theta}$ to $\underset{\sim}{\phi}$ via the transformation

$$\underset{\sim}{\phi} = \tilde{R}(\underset{\sim}{\theta} - \hat{\underset{\sim}{\theta}})$$

If one defines

$$L = \tilde{R}^{-1}$$

then the matrix of scaled second derivative vectors in the $\underset{\sim}{\phi}$ coordinates is related to that of the $\underset{\sim}{\theta}$ coordinates by

$$U_{..} = L^T V_{..} L$$

The $n \times p \times p$ acceleration array $A_{..}$ then becomes

$$A_{..} = Q^T U_{..} = A_{..}^{PE} | A_{..}^{IN}$$

where $A_{..}^{PE}$ is the parameter-effects acceleration array consisting of the first p faces of $A_{..}$ and $A_{..}^{IN}$ is the intrinsic acceleration array consisting of the last $n - p$ faces. Each of the arrays $A_{..}^{IN}$ and $A_{..}^{PE}$ is treated separately. Bates and Watts describe an iterative method for finding the maximum intrinsic curvature IN and the maximum parameter-effects curvature PE. This method has been incorporated into the computer program listed in the Appendix to this book. For each array, the maximum curvature is at least as large as the largest element of the array; thus for a model to behave close to linear, all the elements must be small. Bates and Watts show how the terms a_{ijk} in the array $A_{..}^{PE}$ may be interpreted geometrically; for example, terms of the form a_{iii} represent "scale nonuniformity" due to the ϕ_i parameter, and terms of the form $a_{iji} = a_{iij}$ represent "fanning" of the ϕ_j parameter lines in the ϕ_i direction. Unfortunately, however, these terms do not offer the user any indications of how to find a suitable reparameterization.

Exercises

2.1   Show that expression (2.5) reduces to expression (2.7) for p = 1.

2.2   Show that expression (2.5) reduces to the following for p = 2, where parameters θ_1 and θ_2 are the two elements of $\underset{\sim}{\theta}$, and where it is understood that the terms $\partial g(\underset{\sim}{\theta})/\partial\theta_1$, $\partial g(\underset{\sim}{\theta})/\partial\theta_2$, $\partial^2 g(\underset{\sim}{\theta})/\partial\theta_1^2$, and so on, are to be evaluated at $\hat{\underset{\sim}{\theta}}$:

$$\text{Bias } (\hat{\phi}) = \text{Bias } (\hat{\theta}_1) \frac{\partial g(\underset{\sim}{\theta})}{\partial\theta_1} + \text{Bias } (\hat{\theta}_2) \frac{\partial g(\underset{\sim}{\theta})}{\partial\theta_2}$$

$$+ \frac{1}{2} \text{Var } (\hat{\theta}_1) \frac{\partial^2 g(\underset{\sim}{\theta})}{\partial\theta_1^2} + \frac{1}{2} \text{Var } (\hat{\theta}_2) \frac{\partial^2 g(\underset{\sim}{\theta})}{\partial\theta_2^2}$$

$$+ \text{Cov } (\hat{\theta}_1\hat{\theta}_2) \frac{\partial^2 g(\underset{\sim}{\theta})}{\partial\theta_1 \partial\theta_2}$$

2.3   Show that for $p \geq 2$, the right-hand side of expression (2.5) will contain p terms of the form Bias $(\hat{\theta}_i) \, \partial g(\underset{\sim}{\theta})/\partial\theta_i$, p terms of the form $\frac{1}{2}$ Var $(\hat{\theta}_i) \, \partial^2 g(\underset{\sim}{\theta})/\partial\theta_i^2$, and $p(p-1)/2$ terms of the form Cov $(\hat{\theta}_i\hat{\theta}_j) \, \partial^2 g(\underset{\sim}{\theta})/\partial\theta_i \, \partial\theta_j$, where θ_i and θ_j are elements of the parameter vector $\underset{\sim}{\theta}$ (i, j = 1, 2, ..., p) and where it is understood that all derivatives are to be evaluated at $\hat{\underset{\sim}{\theta}}$.

2.4   Show that expression (2.6) reduces to expression (2.8) for p = 1.

2.5   Show that expression (2.6) reduces to the following for p = 2, where parameters θ_1 and θ_2 are the two elements of $\underset{\sim}{\theta}$ and where it is understood that the derivatives are to be evaluated at $\hat{\underset{\sim}{\theta}}$:

$$\text{Var } (\hat{\phi}) = \text{Var } (\hat{\theta}_1) \left[\frac{\partial g(\underset{\sim}{\theta})}{\partial\theta_1} \right]^2 + \text{Var } (\hat{\theta}_2) \left[\frac{\partial g(\underset{\sim}{\theta})}{\partial\theta_2} \right]^2$$

$$+ 2 \text{ Cov } (\hat{\theta}_1\hat{\theta}_2) \left[\frac{\partial g(\underset{\sim}{\theta})}{\partial\theta_1} \right] \left[\frac{\partial g(\underset{\sim}{\theta})}{\partial\theta_2} \right]$$

2.6   Show that for $p \geq 2$, the right-hand side of expression (2.6) will contain p terms of the form Var $(\hat{\theta}_i) \, [\partial g(\underset{\sim}{\theta})/\partial\theta_i]^2$ and $p(p-1)/2$ terms of the form 2 Cov $(\hat{\theta}_i\hat{\theta}_j) \, [\partial g(\underset{\sim}{\theta})/\partial\theta_i] \, [\partial g(\underset{\sim}{\theta})/\partial\theta_j]$, where θ_i

and θ_j are elements of the parameter vector $\underset{\sim}{\theta}$ (i, j = 1, 2, ...,
p) and where it is understood that all derivatives are to be
evaluated at $\hat{\underset{\sim}{\theta}}$.

2.7 For the same set of data and models considered in the example
illustrated in Sec. 2.7, compute the upper 99% confidence limits
for θ and ϕ. Why are there no lower 99% confidence limits for
these parameters?

2.8 Consider the linear model $E(Y) = \beta X$ in combination with the data
of Table 1.1. Draw the solution locus in sample space (as in
Exercise 1.2). Now generate a pair of ε_t values (ε_1 and ε_2),
corresponding to $X_1 = 2$, $X_2 = 3$, from a normal distribution with
mean zero and variance unity, for $\beta = 3$. Now first add, then
subtract, these ε_t values from βX_t to obtain Y_t^+ and Y_t^- as given
by (2.13) and (2.14), respectively. Determine $\hat{\theta}^+$ and $\hat{\theta}^-$
geometrically by drawing lines from Y_t^+ and Y_t^- perpendicular to
the solution locus and show that the asymmetry measure (2.16) is
zero for this linear model.

2.9 Consider the nonlinear model $E(Y) = X^\theta$ in combination with the
data of Table 1.1, for which the solution locus depicted in Fig.
1.2a is appropriate. Generate a pair of ε_t values corresponding
to $X_1 = 2$, $X_2 = 3$, for $\theta = 2$, and obtain Y_t^+ And Y_t^- values as in
Exercise 2.8. Determine $\hat{\theta}^+$ and $\hat{\theta}^-$ geometrically and show that
the asymmetry measure (2.16) is nonzero for this nonlinear model.

3

Yield-Density Models

3.1 Introduction

In this chapter (and also in Chapters 4-6) we study data sets in combination with various models with a view toward examining their nonlinear behavior in estimation. In particular, we will be attempting to find model functions whose behavior closely approximates that of linear models. The techniques to be employed for examining nonlinearity are described in Chapter 2, the main methods being those of Secs. 2.4-2.6.

The relationship between yield of a crop and the spacing or density of planting is a subject of great interest in agriculture, and a variety of mathematical functions have been proposed to describe the relationship [see the reviews of Willey and Heath (1969) and Mead (1979)]. Essentially two forms of observed response occur in practice, the so-called "asymptotic" and "parabolic" yield-density relations. If X is the plant density in number of plants sown per unit area, and Y is the yield per plant, then W = XY is the total yield per unit area. If W tends to approach an asymptote with increasing plant density it is said to obey the asymptotic yield-density relation, but if W rises to a maximum and then declines at higher densities it is said to obey the parabolic yield-density relation. The majority of crops, including onions, carrots, beans, and tomatoes, follow the asymptotic relation, but others, such as parsnips, sweet corn, and cotton, are distinctly parabolic.

Three three-parameter models will be examined in the present chapter. Two are in wide use, and all three are capable of fitting both asymptotic and parabolic data. Ignoring the stochastic term for the moment, the deterministic components of these models are

$$Y = (\alpha + \beta X)^{-1/\theta} \tag{3.1}$$

$$Y = (\alpha + \beta X + \gamma X^2)^{-1} \tag{3.2}$$

$$Y = (\alpha + \beta X^\phi)^{-1} \tag{3.3}$$

Model (3.1) was proposed by Bleasdale and Nelder (1960), model (3.2) by Holliday (1960), and model (3.3) by Farazdaghi and Harris (1968). When the asymptotic relation applies, so that $\theta = 1$ in (3.1), or $\gamma = 0$ in (3.2), or $\phi = 1$ in (3.3), all three models reduce to the following model of Shinozaki and Kira (1956), which we will term the asymptotic model, although Mead (1979) refers to it as the reciprocal model:

$$Y = (\alpha + \beta X)^{-1} \tag{3.4}$$

A simple biological interpretation for the parameters α and β of (3.4) may be offered. Here the yield per plant Y tends to $1/\alpha$ as the density X tends to zero, so α (or its reciprocal) may be considered to be a measure of the "genetic potential" of a species or variety uninhibited by the effects of competition for environmental resources. At the higher plant densities, as $X \to \infty$, W approaches $1/\beta$ so that β (or its reciprocal) may be considered to be a measure of "environmental potential."

For parabolic yield-density relations, biological interpretation becomes more difficult as α and β may not be able to be separated from the effect of the third parameter. For example, in (3.1) the measure of genetic potential (as $X \to 0$) is $\alpha^{-1/\theta}$ rather than $1/\alpha$. Models (3.2) and (3.3) are much better formulated in this respect, since $Y \to 1/\alpha$ as $X \to 0$, thereby retaining the same interpretation of α (or $1/\alpha$) as for the asymptotic relation (3.4). Nevertheless, β cannot be assumed to measure environmental potential for any of the three parabolic models above, as this parameter cannot be separated from the third parameter as $X \to \infty$.

For obtaining the LS estimates of the parameters of any of the models above, it is more usual to find for given X that log Y, rather than Y, has constant variance (Nelder, 1963). In this chapter, all statistical analyses of model/data set combinations are carried out after logarithmic transformation of models (3.1)-(3.4). For example, model (3.1) for this case of a "multiplicative" stochastic term is

$$E(\log Y) = -\frac{1}{\theta} \log (\alpha + \beta X) \qquad (3.5)$$

3.2 Examination of Nonlinearity in the Yield-Density Models

Table 3.1 presents results of employing the curvature measures of nonlinearity of Bates and Watts (Sec. 2.4) to five sets of data for onions. One of these data sets is for the variety Pukekohe Longkeeper grown in Tasmania (TAS) and was examined previously by Gillis and Ratkowsky (1978). The remaining four sets were supplied by Ian Rogers of the South Australian Department of Agriculture and are for the varieties White Imperial Spanish or Brown Imperial Spanish grown in four South Australian locations: Mt. Gambier (MG), Uraidla (U), Purnong Landing (PL), and Virginia (V). These data are given in Appendix 3.A

Table 3.1 Intrinsic (IN) and Parameter-Effects (PE) Curvatures for Yield-Density Model/Data Set Combinations for Onion Spacing Trials in Tasmania and Four South Australian Localities

Locality	Curvature component	Bleasdale-Nelder (3.1)	Holliday (3.2)	Farazdaghi-Harris (3.3)	Asymptotic (3.4)
TAS	IN	.15	.04	.10	.03
	PE	23.54	.10	13.47	.06
MG	IN	.27	.04	.16	.03
	PE	335.19	.08	61.75	.03
U	IN	.35	.05	.19	.03
	PE	334.84	.07	86.81	.05
PL	IN	.15	.03	.09	.02
	PE	84.62	.05	37.18	.03
V	IN	.22	.04	.12	.03
	PE	133.76	.07	38.22	.04

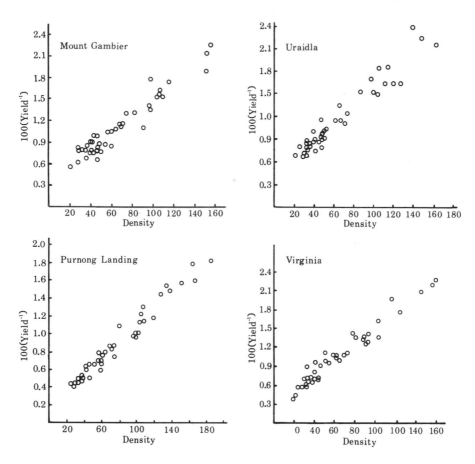

Fig. 3.1 Graph of four data sets (Appendix 3.A) on onions from South Australia.

and are shown graphically in Fig. 3.1. The LS estimates of the parameters for all models studied are given in Appendix 3.B. The sample size for the Tasmanian data set is 20, whereas the four South Australian data sets are each of size 42. The critical values $1/(2\sqrt{F})$ associated with the curvature measures for assessing whether the nonlinearity is significant at level .05 are, for the three-parameter parabolic models: .28 for TAS and .30 for the other four data sets; for the two-parameter asymptotic model: .27 for TAS and .28 for the other four data sets.

Several conclusions are immediately apparent from the results in Table 3.1. The asymptotic model (3.4) behaves very much like a linear model, as does the Holliday model (3.2). Both the intrinsic (IN) and parameter-effects (PE) curvatures are, for both models, far from the critical values; the presence of the extra term γX^2 in (3.2) results only in a very small increase in the nonlinearity when compared with (3.4). For the Farazdaghi-Harris model (3.3), the presence of the additional parameter ϕ increases the intrinsic nonlinearity somewhat, although not to the point of reaching the nominal 5% significance level, but markedly increases the parameter-effects nonlinearity to unacceptable levels. The results for the Bleasdale-Nelder model (3.1) are even more extreme; in addition to exceptionally large parameter-effects curvatures, the intrinsic curvatures are also close to the critical values, exceeding the critical value for one data set.

An idea of which estimators are behaving nonlinearly in (3.1) and (3.3) is obtained by examining the results of Box's bias calculation (Sec. 2.5). These are given in Table 3.2, with bias expressed as a percentage of the parameter estimate. For model (3.1), the LS estimates of α have a very considerable bias (sometimes exceeding 100%), with the bias in β also being large although of a somewhat lesser magnitude. The situation is the reverse for model (3.3), where the bias in the estimates of β is considerable, whereas the bias in the estimates of α is very much smaller. For (3.3), the bias in $\hat{\alpha}$ is negative in all five data sets, that is, α tends to be consistently underestimated.

Table 3.2 Bias in Parameter Estimates of Models (3.1) and (3.3) Expressed as a Percentage of the Estimate

Bleasdale-Nelder (3.1)					Farazdaghi-Harris (3.3)				
Parameter	TAS	MG	U	PL	V	Parameter	TAS	MG	U	PL	V
α	32	152	257	59	168	α	-1.1	-2.6	-6.2	-2.6	-8.5
β	8	28	55	15	52	β	8	58	104	20	26
θ	-.1	.4	-.1	0	-.1	ϕ	.2	.5	.9	.2	.3

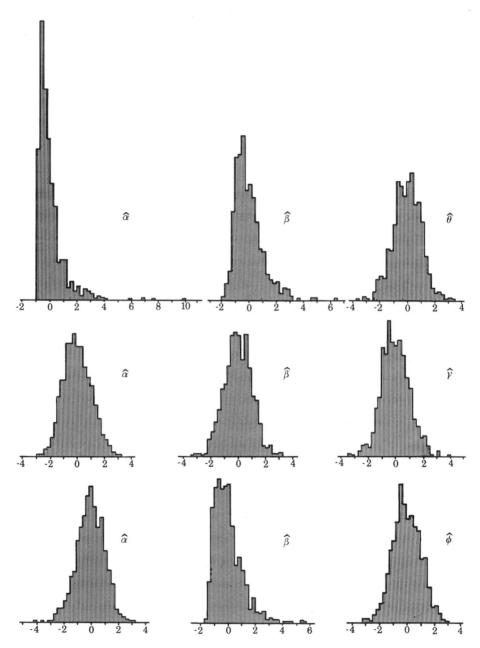

Fig. 3.2 Histogram of results for yield-density models. Top row: $\hat{\alpha}$, $\hat{\beta}$, and $\hat{\theta}$ for Bleasdale–Nelder model (3.1); middle row: $\hat{\alpha}$, $\hat{\beta}$, and $\hat{\gamma}$ for Holliday model (3.2); bottom row: $\hat{\alpha}$, $\hat{\beta}$, and $\hat{\phi}$ for Farazdaghi–Harris model (3.3).

The estimates of both θ in (3.1) and φ in (3.3) are almost unbiased, an interesting result, especially as it is the presence of these parameters in the models that causes the model to behave nonlinearly.

Simulation studies were carried out in the manner described in Sec. 2.6, with 1000 sets of simulated data being generated for each model. Typical results are presented in Fig. 3.2, corresponding to the Tasmanian data set (TAS). This is the same data set that was used in the simulation studies presented in Gillis and Ratkowsky (1978) and the reader is referred to that paper for further details.

For model (3.1) the histograms of Fig. 3.2 show a strong positive skewness for the estimates of α and a somewhat smaller positive skewness for the estimates of β, whereas the estimates of θ follow a normal distribution closely. For model (3.2) the estimates of all parameters appear to follow a normal distribution closely. For model (3.3) a strong positive skewness is apparent for the estimates of β, and a smaller, but still quite perceptible, negative skewness for the estimates of α. The estimates of φ appear to be consistent with that of a normal distribution.

Using the estimates depicted in Fig. 3.2 one may carry out the statistical tests described in Sec. 2.6. Table 3.3 presents results of these tests for the models (3.1)-(3.3), which confirm the visual impression conveyed by the histograms in Fig. 3.2. The LS estimators of all parameters of the Holliday model (3.2) exhibit close-to-linear behavior as defined in Sec. 1.4, whereas the estimates of α and β in (3.1) and of β in (3.3) exhibit a high degree of positive bias, excess variance, positive skewness, and excess kurtosis. For the estimates of α in (3.3), the results in Table 3.3 quantify the pictorial impression of negative skewness gained from Fig. 3.2.

A further simulation study was carried out as described in Sec. 2.9 to estimate the asymmetry measure of bias for the Bleasdale-Nelder model in combination with this TAS data set. Percentage biases, based on the asymmetry measure (2.16), were 36, 8, and -.2% for the estimators of α, β, and θ, respectively, which compare well with the bias values in Tables 3.2 and 3.3 for these same parameters. The correlation

Table 3.3 Results from Simulation Studies of Models (3.1), (3.2), and (3.3) for the TAS Data Set

Parameter	Model	Percent bias	Percent excess variance	Coefficient of skewness	Coefficient of excess kurtosis
α	(3.1)	38.6^{**}	165.6^{**}	3.36^{**}	19.43^{**}
α	(3.2)	$.25^{ns}$	1.11^{ns}	$.10^{ns}$	$-.25^{ns}$
α	(3.3)	-1.45^{*}	2.51^{ns}	$-.28^{**}$	$.22^{ns}$
β	(3.1)	8.43^{**}	30.6^{**}	1.37^{**}	3.40^{**}
β	(3.2)	$-.04^{ns}$	3.08^{ns}	$-.05^{ns}$	$.10^{ns}$
β	(3.3)	9.39^{**}	35.08^{**}	1.45^{**}	3.63^{**}
θ	(3.1)	$-.09^{ns}$	3.76^{ns}	$-.08^{ns}$	$.24^{ns}$
γ	(3.2)	$-^{a}$	3.83^{ns}	$.10^{ns}$	$.29^{ns}$
ϕ	(3.3)	$.10^{ns}$	1.70^{ns}	$-.04^{ns}$	$-.16^{ns}$

Note: $^{ns}P > .05$; $^{*}P < .05$; $^{**}P < .01$.

[a] The true value of γ in (3.2) is zero, which would result in an infinite percent bias; hence it is omitted from the table.

coefficients using (2.18) were $-.545$, $-.854$, and $-.974$, respectively, which correlates well with the fact that the nonlinearity in the estimators is greatest for $\hat{\alpha}$, second greatest for $\hat{\beta}$, and low for $\hat{\theta}$.

3.3 Choice of Yield-Density Model

The results presented in Sec. 3.2 indicate the suitability of the Holliday model (3.2) for yield-density studies. Although these results are for onions, similar results for this model were obtained for other crops. For example, Gillis and Ratkowsky (1978) presented results for a total of seven crops, three of which exhibit parabolic yield-density behavior. The curvature measures of nonlinearity of Bates and Watts (1980) were not published at that time and therefore were not used in that study. However, the results obtained from Box's bias calculation and the simulation studies given in Table 3 of that paper clearly show

that the Holliday model behaves in a manner closely resembling the
behavior of a linear model, whereas the Bleasdale-Nelder model (3.1)
does not. In view of the expressed aim in this book of favoring models
that behave like linear models, model (3.2) appears to be entirely
suitable.

The question of goodness of fit, mentioned in Section 2.3, now
arises. Does the Holliday model fit data well enough, compared with
the alternative models (3.1) and (3.3), to justify the claim that it
is suitable for general use in yield-density studies? Peter Gillis,
formerly of the Tasmanian Department of Agriculture, has made an
extensive (unpublished) study of all yield-density data that he could
find in the literature. In addition to the large number of crops that
exhibit asymptotic behavior, he found data for several crops which
manifest parabolic behavior, such as parsnips and maize. In general,
the three models (3.1)-(3.3) tend to fit these data about equally as
well, as determined by an examination of the residual variance and of
the pattern of the residuals using runs tests. Each of these models
had the smallest residual variance about equally often, and in
virtually all cases these variances were very close to each other. As
there appears to be no clear choice between models on the basis of
goodness of fit, the decision as to which model to use can rest solely
on the statistical properties of the estimators. In this regard, model
(3.2) must clearly be favored. Mead (1979) has argued that model (3.1)
be retained for its "biological advantages," but does not enumerate
what these advantages are. He points out, correctly, that if θ is fixed
to have a constant value for a given crop, the LS estimators of α and
β then behave close to linear, as is the case, for example, for the
results using model (3.4), which is equivalent to $\theta = 1$. He advocates
searching for an invariant value of θ over a number of sets of data,
and then employing model (3.1) with θ fixed at that value. If biological
advantages could indeed be demonstrated for model (3.1) in preference
to model (3.2), this suggestion could have merit, but until that time
the use of model (3.2) seems to be indicated.

Appendix 3.A: Onion Data for Yield Y (g/plant) Versus Density X (Plants/m^2) from Four South Australian Localities

Brown Imperial Spanish				White Imperial Spanish			
Mount Gambier (MG)		Uraidla (U)		Purnong Landing (PL)		Virginia (V)	
X	Y	X	Y	X	Y	X	Y
20.64	176.58	22.30	148.57	23.48	223.02	18.78	272.15
26.91	159.07	25.86	125.30	26.22	234.24	21.25	235.23
26.91	122.41	29.09	150.69	27.79	221.68	23.23	180.47
28.02	128.32	29.74	147.42	32.88	221.94	27.18	177.31
32.44	125.77	31.68	117.10	33.27	197.45	30.15	141.28
34.28	126.81	31.68	116.64	36.79	189.64	31.63	169.39
35.76	147.77	32.00	129.66	37.58	211.20	32.12	138.17
36.49	117.29	32.32	131.54	37.58	191.36	32.62	171.81
38.71	133.49	32.32	151.50	41.49	156.62	32.62	112.02
39.44	128.87	34.91	121.80	42.66	168.12	33.61	156.09
39.81	110.04	35.23	125.67	44.23	197.89	37.07	137.29
40.92	111.15	38.47	117.78	44.23	154.14	38.55	154.10
42.76	134.12	39.44	101.50	51.67	153.26	39.54	124.17
43.50	99.94	41.05	113.22	55.58	142.79	39.54	146.28
45.34	128.70	41.70	136.43	55.58	126.17	41.02	105.47
45.71	152.17	44.28	117.54	57.93	167.95	42.50	139.24
46.82	100.36	45.90	87.20	58.71	144.54	43.98	148.31
47.18	123.32	46.55	107.41	59.50	151.30	45.47	110.44
47.92	114.44	48.16	129.68	60.67	130.52	49.92	90.72
48.66	131.27	48.49	104.63	62.63	125.30	50.90	102.61
53.45	115.12	48.81	114.15	67.71	114.05	53.87	107.36
55.66	95.52	49.78	99.85	70.06	116.31	57.82	92.66
59.35	94.94	50.43	111.65	70.45	120.71	61.78	96.52
59.72	119.28	51.72	98.09	73.98	134.16	61.78	94.71
63.04	93.64	61.42	87.85	73.98	114.48	63.75	99.86
67.09	85.73	65.29	75.45	78.67	91.17	67.71	93.37
68.93	89.26	67.23	87.01	95.90	101.27	71.66	89.78
69.30	88.55	71.44	90.10	96.68	97.33	77.59	69.34
73.36	76.81	73.05	81.08	96.68	101.37	80.56	73.74
80.73	76.63	86.63	65.33	101.38	97.20	86.49	75.17
89.58	90.53	96.00	58.49	103.72	87.12	88.46	72.98
95.47	71.28	98.91	65.67	104.51	81.71	89.45	79.94
98.05	56.61	103.44	67.19	105.68	76.44	90.93	79.13
98.42	75.09	105.05	54.01	108.03	87.10	92.91	70.93
102.48	65.26	111.19	60.92	117.82	84.54	101.81	60.99
105.80	64.48	113.78	53.48	127.21	69.09	103.78	74.09
106.53	61.84	119.92	61.62	134.26	64.40	115.15	49.45
108.75	65.19	120.89[a]	26.32	137.39	66.81	123.06	56.65
115.38	57.10	126.71	61.21	151.87	63.01	144.31	47.84
150.77	52.68	138.99	41.67	163.61	55.45	155.68	40.03
152.24	47.01	146.75	45.26	166.35	62.54	158.15	38.70
155.19	44.28	160.97	46.45	184.75	54.68	180.39[a]	28.96

[a] These points are suspected outliers and are omitted from all calculations presented in Chapter 7.

Source: Courtesy of I.S. Rogers of the South Australian Department of Agriculture and Fisheries.

Appendix 3.B: LS Estimates of the Parameters and Residual Variances in Yield-Density Models Assuming Var (log Y) Constant for Data Sets of Appendix 3.A

Locality	Parameter	Bleasdale-Nelder	Holliday	Farazdaghi-Harris	Asymptotic
MG	α	.03267	.004524	.005043	.003739
	β	$.3581(10^{-3})$	$.08113(10^{-3})$	$.02869(10^{-3})$	$.1093(10^{-3})$
	θ, γ, or ϕ	.6346	$.1976(10^{-6})$	1.263	
	σ^2	.01229	.01231	.01226	.01242
U	α	.02019	.004220	.004859	.003462
	β	$.3222(10^{-3})$	$.1012(10^{-3})$	$.03877(10^{-3})$	$.1291(10^{-3})$
	θ, γ, or ϕ	.7162	$.1933(10^{-6})$	1.235	
	σ^2	.02303	.02308	.02295	.02280
PL	α	.003842	.002054	.002294	.001883
	β	$.1302(10^{-3})$	$.08571(10^{-3})$	$.06032(10^{-3})$	$.09159(10^{-3})$
	θ, γ, or ϕ	.9055	$.03808(10^{-6})$	1.080	
	σ^2	.007265	.007287	.007253	.007153
V	α	.001803	.002084	.001648	.001809
	β	$.1418(10^{-3})$	$.1311(10^{-3})$	$.1581(10^{-3})$	$.1420(10^{-3})$
	θ, γ, or ϕ	1.000	$.07796(10^{-6})$	.9791	
	σ^2	.01568	.01558	.01568	.01529

Exercises

3.1 Determine the LS estimates of the parameters of models (3.1)-(3.4), assuming that log Y has constant variance, for the Uraidla and Virginia data sets in Appendix 3.A, omitting the suspected outliers. Compare your results with the estimates tabulated in Appendix 3.B for these same data sets with the suspected outliers included. Are there any major differences?

3.2 Using the LS estimates of the parameters and the estimated residual variance for each of the models examined in Exercise 3.1, calculate the curvature measures of nonlinearity of Bates and Watts and Box's bias calculation for the Uraidla and Virginia data sets in Appendix 3.A, omitting the suspected outliers. Compare your results with the results in Table 3.1 and Table 3.2 [for models (3.1) and (3.3)] for these same data sets with the suspected outliers included. Are there any major differences?

3.3 Using the results from Exercise 3.1, carry out tests of significance for the parameters θ, γ, and ϕ and show that these values do not differ significantly from $\theta = 1$, $\gamma = 0$, and $\phi = 1$, respectively. In other words, show that the asymptotic yield–density model given by expression (3.4) is an adequate model for these data. Enterprising readers may wish to carry out similar calculations for the Mount Gambier and Purnong Landing data sets, which also show that the asymptotic model is adequate.

3.4   Carry out simulation studies, consisting of 1000 trials each, for models (3.1)–(3.3) assuming that log Y has constant variance, in combination with each of four data sets in Appendix 3.A, generating the data using the parameter estimates and residual variances given in Appendix 3.B. To ensure convergence to the LS esimates of the parameters for each of the 1000 trials, good initial parameter estimates are required and it will be necessary to use the methods described in Sec. 8.2 for models (3.1) and (3.3). For model (3.1), use the procedure described in Sec. 8.2.3, and for model (3.3), use the weighted least–squares criterion which results in equations (8.19) of Sec. 8.2.2. Present your results for each data set in tables similar to that of Table 3.3, after carrying out tests of significance for the bias, excess variance, skewness, and kurtosis using the methods described and detailed in Sec. 2.6. Also display your results for each parameter in the form of histograms similar to those of Fig. 3.2 for the TAS data set and discuss whether your histograms allow similar conclusions to be drawn. Compare the percent bias from your simulation studies with the values tabulated in Table 3.2 for models (3.1) and (3.3) determined using Box's calculation. Do they agree well as to order of magnitude? Readers should note the agreement in order of magnitude between the percent bias from the simulation study on the TAS data set tabulated in Table 3.3 and the calculated percent bias values for that data set in Table 3.2.

4

Sigmoidal Growth Models

4.1 Introduction

Processes producing sigmoidal or "S-shaped" growth curves are widespread in biology, agriculture, engineering, and economics. Such curves start at some fixed point and increase their growth rate monotonically to reach an inflection point; after this the growth rate decreases to approach asymptotically some final value. Figure 4.1 shows four data sets obtained from a variety of vegetative growth process, all of which exhibit sigmoidal behavior. The range of sigmoidal vegetative growth is vast, including microscopic processes such as the growth of bean root cells as in Fig. 4.1, and the growth of whole forests (Yang et al., 1978).

Numerous mathematical functions have been proposed for modeling sigmoidal curves, many of which are claimed to have some underlying theoretical basis. Among these are the Gompertz, the logistic, the Richards (1959), the Morgan-Mercer-Flodin (1975), and a model derived from the Weibull (1951) distribution, here designated as a Weibull-type model. The following forms of these models are considered initially (only the deterministic component is shown), although later all these models will be considered in various other model functions:

$$\text{Gompertz:} \qquad Y = \alpha \exp \left[-\exp (\beta - \gamma X) \right] \qquad (4.1)$$

$$\text{Logistic:} \qquad Y = \frac{\alpha}{1 + \exp (\beta - \gamma X)} \qquad (4.2)$$

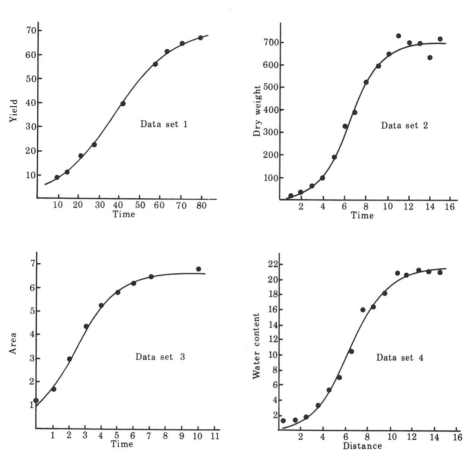

Fig. 4.1 Graph of data on four vegetative growth processes. Data set 1: pasture regrowth, yield versus time (data courtesy of W. F. Hunt). Data set 2: onion bulbs plus tops, yield versus time. Data set 3: cucumber cotyledons, area versus time (data from Gregory, 1956). Data set 4: bean root cells, water content versus distance from growing tip (data from Heyes and Brown, 1956).

Richards:

$$Y = \frac{\alpha}{[1 + \exp{(\beta - \gamma X)}]^{1/\delta}} \quad (4.3)$$

Morgan–Mercer–Flodin (MMF): $Y = \dfrac{\beta\gamma + \alpha X^{\delta}}{\gamma + X^{\delta}}$ (4.4)

Weibull type: $\qquad Y = \alpha - \beta \exp(-\gamma X^{\delta})$ $\qquad$ (4.5)

The four sets of data drawn in Fig. 4.1 are also presented in Appendix 4.A. These data sets have been chosen for consideration in this book as each of them covers virtually the full range of the response variable Y and each consists of measurements on *independent* experimental units, thereby avoiding the added statistical complexity associated with the correlated errors that arise when repeated measurements are made on the same individuals (e.g., see Sandland and McGilchrist, 1979). The four data sets are typical of ones found in practice with respect to sample size, which vary from nine for pasture regrowth and cucumber cotyledons to 15 for onion bulbs plus tops and bean root cells. As stressed in Sec. 2.3, it is very difficult to come to any decision about which of several competing models fits the data better when small sample sizes are involved. This is particularly true for the four data sets considered here, as graphs of residuals, and runs tests performed thereon, give no indication of which of models (4.1)-(4.5) is preferable from the point of view of goodness of fit. In addition, inspection of Fig. 4.1 does not indicate any clear-cut choice between assuming an additive error term or a multiplicative error term. Hence all growth models were fitted in two ways, as detailed in the following section.

4.2 Stability of Parameter Estimates to Varying Assumptions About the Error Term

The first series of estimations are carried out assuming an additive error term, which means that models (4.1)-(4.5) are of the form

$$Y_{t_A} = f(X_t, \underset{\sim}{\theta}) + \varepsilon_{t_A}$$

where $\underset{\sim}{\theta}$ designates the vector of the parameters α, β, and γ (and δ where appropriate) to be estimated, and ε_{t_A} is assumed to be iidN with mean zero and unknown variance σ_A^2. The second series of estimations are carried out assuming a multiplicative error term, which means that models (4.1)-(4.5) are logarithmically transformed and are of the form

$$\log Y_{t_M} = \log f(X_t, \underset{\sim}{\theta}) + \varepsilon_{t_M}$$

where ε_{t_M} is assumed to be iidN with mean zero and unknown variance σ_M^2. As always throughout this book, the parameter estimates are obtained by least squares, by minimizing $\sum \varepsilon_{t_A}^2$, on the one hand, and $\sum \varepsilon_{t_M}^2$, on the other hand, using the Gauss–Newton method.

Table 4.1 gives the LS estimates of the parameters of (4.1)–(4.5) for each of the four data sets, for both the additive and multiplicative error assumptions. For the three-parameter models (4.1) and (4.2), the logistic model always has a lower residual variance than the Gompertz model. In addition, the logistic is also more stable with respect to the two different assumptions about the stochastic term, as the LS estimates are quite close to each other for both the additive and multiplicative error assumptions. Although the stability of the LS estimates to varying assumptions about the stochastic term was not one of the suggested criteria for model assessment discussed in Chapter 2, it is clearly desirable that the parameter estimates of a model remain stable in situations where it is difficult or impossible to decide the distributional properties of the stochastic term. For the four-parameter models (4.3)–(4.5), these three models have residual variances of comparable order of magnitude and thereby fit the data about equally as well (except in the case of the MMF model for the cucumber cotyledon data set, where the fit is exceptionally good, possibly owing simply to chance). With regard to the stability of the LS estimates to the two different assumptions about the stochastic term, (1) for the Richards model, the estimates undergo considerable variation for all parameters except α; (2) for the MMF model, the estimates for parameter γ vary enormously and those for parameter δ exhibit moderately high variation; (3) for the Weibull-type model the estimates are relatively stable for virtually all parameters except γ. It will be seen from results to be presented in Sec. 4.3 that there is a close connection between the stability of the parameter estimates and the bias in those estimates; the parameters whose estimates have the least bias are also the most stable.

Table 4.1 Parameter Estimates for Growth Models, Additive and Multiplicative Errors

| | | Three-parameter models | | | | Four-parameter models | | | | | |
| | | Gompertz (4.1) | | Logistic (4.2) | | Richards (4.3) | | Morgan-Mercer-Flodin (MMF) (4.4) | | Weibull type (4.5) | |
Data set	Parameter	Add.	Mult.	Add.	Mult.	Add.	Mult.	Add.	Mult.	Add.	Mult.
Pasture regrowth	α	82.83	96.94	72.46	73.20	69.62	72.53	80.96	93.26	69.96	73.48
	β	1.224	1.159	2.618	2.591	4.255	2.773	8.895	7.328	61.68	66.41
	γ	.037	.030	.067	.066	.089	.068	49,577	8,773	.000100	.000261
	δ					1.724	1.072	2.828	2.298	2.378	2.106
	σ^2	3.63	.0047	1.34	.0032	1.21	.0038	2.71	.0042	1.68	.0038
Onion bulbs plus tops	α	723.1	903.1	702.9	703.3	699.6	702.5	723.9	827.8	695.0	708.9
	β	2.500	1.682	4.443	4.395	5.277	4.449	33.35	15.94	673.5	693.9
	γ	.450	.254	.689	.680	.760	.686	6,266	550	.00152	.00293
	δ					1.279	1.013	4.641	3.178	3.262	2.891
	σ^2	1,134	.0215	744	.0047	799	.0051	1,015	.0130	712	.0083
Cucumber cotyledons	α	6.925	7.585	6.687	6.894	6.684	6.497	6.986	6.836	6.656	6.478
	β	.768	.666	1.745	1.664	1.780	4.832	1.181	1.216	5.549	5.280
	γ	.493	.366	.755	.669	.759	1.293	12.96	15.53	.118	.0826
	δ					1.017	2.772	2.475	2.707	1.763	2.138
	σ^2	.0619	.0177	.0353	.0104	.0424	.0092	.0048	.0004	.0268	.0026
Bean root cells	α	22.51	36.20	21.51	23.34	21.20	20.78	22.08	23.95	21.10	21.22
	β	2.106	1.391	3.957	3.398	5.691	9.939	1.653	1.264	19.81	19.99
	γ	.388	.170	.622	.484	.777	1.219	5,586	854.8	.00177	.00218
	δ					1.619	3.111	4.560	3.470	3.180	3.062
	σ^2	1.049	.0440	.518	.0193	.502	.0128	.579	.0054	.495	.0040

4.3 Examination of Nonlinearity in the Sigmoidal Growth Models

Table 4.2 gives the results of employing the curvature measures of
nonlinearity of Bates and Watts (Sec. 2.4) to the four data sets for
the additive error assumption only. A similar picture generally
emerges with the multiplicative error assumption. The critical values
$1/(2\sqrt{F})$ associated with the curvature measures for assessing whether
the nonlinearity is significant at level .05 are, for the
three-parameter models: .229 (pasture, cucumber) and .268 (onion,
bean); for the four-parameter models: .219 (pasture, cucumber) and .273
(onion, bean). The intrinsic curvatures in Table 4.2 are less than
these critical values for the Gompertz, logistic, MMF and Weibull
models for each of the four data sets. In contrast, the intrinsic
curvatures of the Richards model exceed the critical values for each
data set, indicating that the solution locus departs significantly from
linearity for this model with these data sets. With respect to the
parameter-effects curvatures in Table 4.2, those for the logistic model
show the least departure from the limiting linear case of "uniform
coordinates," although the nonlinearity is significant for each data
set. All four-parameter models exhibit high parameter-effects
curvature.

Table 4.2   Estimated Intrinsic (IN) and Parameter-Effects (PE)
Curvatures for Growth Models: Additive Error Assumption

Data set	Curvature component	Gompertz (4.1)	Logistic (4.2)	Richards (4.3)	MMF (4.4)	Weibull (4.5)
Pasture	IN	.090	.073	.267	.180	.130
	PE	2.324	.644	6.679	90.970	42.675
Onion	IN	.234	.131	.330	.257	.271
	PE	.700	.379	6.271	31.319	16.371
Cucumber	IN	.121	.118	.332	.103	.188
	PE	.633	.351	14.811	1.154	1.878
Bean	IN	.232	.107	.295	.210	.232
	PE	.880	.372	4.268	24.934	13.253

   A high parameter-effects curvature indicates that at least one
parameter in the model is departing strongly from linear behavior, and
the calculated bias using the method of Box (Sec. 2.5), presented in
Table 4.3 as a percentage of the estimate, may help indicate which
parameter or parameters are responsible. For the logistic model, the
bias values in Table 4.3 are always less than 1%, suggesting that the
nonlinear behavior of this model may be small in practical terms.
However, there is no guarantee that a small percentage bias will
necessarily indicate close-to-linear behavior of the model in
estimation (see Sec. 9.5). It is often a useful guide, but only a
simulation study such as described in Sec. 2.6 can fully settle the
question. For the Gompertz model, the biases are all somewhat higher
than for the logistic model, and this is consistent with its higher
parameter-effects curvature. However, the highest bias is only
slightly in excess of 1%.

Table 4.3   Box's Percentage Bias in LS Estimates: Additive Error
Assumption

Data set	Parameter	Gompertz (4.1)	Logistic (4.2)	Richards (4.3)	MMF (4.4)	Weibull (4.5)
Pasture	α	.894	.135	.376	1.525	.490
	β	.602	.172	2.305	−1.645	.686
	γ	.428	.158	2.447	119.478	29.847
	δ			3.567	.921	.332
Onion	α	.185	.072	.221	.313	.104
	β	1.088	.616	5.135	−2.483	.186
	γ	1.007	.625	5.250	86.600	16.379
	δ			8.796	1.424	.916
Cucumber	α	.185	.065	.209	.040	.112
	β	.798	.394	1.085	−.047	.156
	γ	.659	.415	5.108	1.272	.348
	δ			11.599	.172	.757
Bean	α	.270	.071	.183	.237	.103
	β	1.066	.438	4.431	−.794	.157
	γ	.976	.444	4.232	55.341	11.485
	δ			6.363	.942	.697

For the Richards model, moderately high biases, always exceeding 1%, are obtained for the LS estimators of three of the four parameters. Recall that it was these same three parameters whose LS estimators showed instability to the varying assumptions about the stochastic term. This highlights the close connection between instability of estimators and nonlinear behavior of the model. For the MMF model, the LS estimators of parameter γ have the greatest bias, this parameter being mostly responsible for the high parameter-effects curvature in Table 4.2. For the Weibull-type model, the LS estimators of all parameters except γ have percentage biases of less than 1%. We recall that it was parameter γ whose estimators were the most unstable to the two differing assumptions about the stochastic term.

The results of this section can be summarized by noting that in the case of the three-parameter models, both the Gompertz and logistic models have sufficiently low intrinsic curvatures such that the "planar approximation" (actually a three-dimensional hyperplane) to the solution locus is an acceptable one. Both models exhibit significant parameter-effects nonlinearities, but that for the logistic model was not serious in practical terms, as the results of simulation studies (see Exercise 4.1) show that the distribution of the estimates for each parameter is very close to that expected for the normal distribution. The Gompertz model seems less suitable than the logistic model, as the departure from linear behavior is greater for all parameters. For the four-parameter models, the MMF and Weibull-type models have acceptably low intrinsic nonlinearities. The Richards model, on the other hand, has a significant curvature in its solution locus, although the departure from linearity is not excessive. For the MMF and Weibull-type models, most of the parameter-effects nonlinearity is centered in a single parameter (γ in both models), suggesting that much of the nonlinearity can be removed by a suitable reparameterization involving this parameter. The situation with the Richards model is more difficult as all but one of its parameters are contributing substantially to the overall nonlinear behavior. The next section is devoted to a study of reparameterization of the five models.

4.4 Searching for Better Parameterizations of the Growth Models

The five models studied in this chapter were considered initially in the model functions given by (4.1)–(4.5). We now consider other model functions for each model, the object being to find ones which have smaller parameter-effects nonlinearity, whose behavior may thereby more closely approach that of a linear model. We will employ the curvature measure of parameter-effects nonlinearity (Sec. 2.4), the bias calculation of Box (Sec. 2.5) and simulation studies (Sec. 2.6). Each model will be considered in turn in the following subsections.

4.4.1 Gompertz model

In addition to the model function initially considered, namely (4.1),

$$Y = \alpha \exp \left[-\exp \left(\beta - \gamma X \right) \right]$$

another form in which the Gompertz model is commonly used is

$$Y = \exp \left(\alpha - \beta \gamma^X \right) \tag{4.6}$$

Table 4.4 presents results for the curvature measure of parameter-effects nonlinearity and for the Box bias calculation for these model functions assuming additive error. The results for model function (4.1) have already been given (Tables 4.2 and 4.3) but are presented again in Table 4.4 to facilitate comparison with model function (4.6). The intrinsic curvatures are the same for both model functions, as reparameterization does not alter the position of the solution locus (see Exercise 1.3). Reparameterization does not change the parameter-effects curvatures in any consistent way, decreasing it for two of the data sets, but increasing it for the other two data sets. The percentage biases in the parameter estimates are consistent, however; those for parameters α and γ always decrease in magnitude with reparameterization, whereas those for β always increase, the latter exceeding 1% in each data set.

Simulation studies confirm that parameter β is the most nonlinear-behaving parameter in model function (4.6), as the estimates of β exhibit marked departures from a normal distribution (see Exercise

Table 4.4 Gompertz Model: Box's Percentage Bias and Intrinsic (IN) and Parameter-Effects (PE) Curvatures for Model Functions (4.1) and (4.6): Additive Error Assumption

Data set	Parameter	Model function (4.1)	Model function (4.6)
Pasture regrowth	α	.894%	.149%
	β	.602%	1.019%
	γ	.428%	-.015%
	IN	.090	.090
	PE	2.324	1.644
Onion bulbs plus tops	α	.185%	.021%
	β	1.088%	6.773%
	γ	1.007%	-.319%
	IN	.234	.234
	PE	.700	2.629
Cucumber cotyledons	α	.185%	.070%
	β	.798%	1.039%
	γ	.659%	-.192%
	IN	.121	.121
	PE	.633	.442
Bean root cells	α	.270%	.065%
	β	1.066%	5.009%
	γ	.976%	-.273%
	IN	.232	.232
	PE	.880	1.983

4.3). However, different results are obtained using the multiplicative error assumption. Box's bias calculation predicts biases in excess of 1% in the estimates of α for all four data sets (e.g., in the bean root cells data set the predicted bias exceeds 8%), and simulation studies confirm that the LS estimators of this parameter, which represents the asymptote, are significantly biased and non-normal in distribution (see Exercise 4.4). Thus the nature of the stochastic term may lead to different conclusions about what constitutes a suitable parameterization. Readers may wish to search for other parameterizations as an exercise, but we recall from an examination of the residual variance in Table 4.1 that the Gompertz model, whether one considers additive error or multiplicative error, was the worst fitting model of the five models for each data set. Hence the Gompertz model, which is widely used in econometric modeling, is perhaps not a very good model for vegetative growth processes, if the data sets considered in this chapter are typical.

4.4.2 Logistic model

The logistic model, studied initially in model function (4.2),

$$Y = \frac{\alpha}{1 + \exp\ (\beta\ -\ \gamma X)}$$

consistently had low parameter-effects nonlinearity. Nevertheless, it is of interest to see whether there are any other parameterizations which may be even more suitable than (4.2) for practical use. The following model functions of the logistic model are considered:

$$Y = \frac{1}{\alpha + \beta\ \exp\ (-\gamma X)} \tag{4.7}$$

$$Y = \frac{1}{\alpha + \beta\gamma^X} \tag{4.8}$$

$$Y = \frac{\alpha}{1 + \exp\ (\beta)\gamma^X} \tag{4.9}$$

$$Y = \frac{1}{\alpha + \exp\ (\beta)\gamma^X} \tag{4.10}$$

$$Y = \frac{\alpha}{1 + \beta\ \exp\ (-\gamma X)} \tag{4.11}$$

Model function (4.11) was considered by Oliver (1966).

Results are presented in Table 4.5, for the additive error assumption, for the parameter-effects curvature measure and for the percentage bias in the estimators of α, β, and γ. As the intrinsic curvature is unaltered by reparameterization, it is omitted from Table 4.5. Model functions (4.7), (4.8), and (4.11) have relatively high parameter-effects curvatures for two of the data sets, which is also reflected in the high percentage bias in the estimators of β, being of the order of 9% for onions and 5% for beans. It is difficult to choose between model functions (4.2), (4.9), and (4.10) on the basis of parameter-effects curvature alone, but the fact that the percentage bias in the estimators of β for model function (4.10) is greater than 1% for three of the four data sets tends to make one exclude that model function in favor of either (4.2) or (4.9) provided that we have an equal interest in all parameters.

Simulation studies confirm the close connection between percentage bias and linear behavior of the LS estimators for these models (see Exercise 4.5). Those parameter estimators for which the percentage biases in Table 4.5 exceed 1% all have distributions which differ significantly from the normal distribution, as judged by the

Table 4.5 Box's Percentage Bias and Parameter-Effects (PE) Curvature for Various Parameterizations of the Logistic Model: Additive Error Assumption

Data set	Parameter	Model function					
		(4.2)	(4.7)	(4.8)	(4.9)	(4.10)	(4.11)
Pasture	α	.135%	−.078%	−.078%	.135%	−.078%	.135%
regrowth	β	.172%	.859%	.859%	.172%	−.207%	.840%
	γ	.158%	.158%	−.010%	−.010%	−.010%	.158%
	PE	.644	.508	.521	.624	.448	.582
Onion bulbs	α	.072%	−.032%	−.032%	.072%	−.032%	.072%
plus tops	β	.616%	9.138%	9.138%	.616%	−1.270%	8.889%
	γ	.625%	.625%	−.265%	−.265%	−.265%	.625%
	PE	.379	2.836	3.264	.462	.464	2.704
Cucumber	α	.065%	−.027%	−.027%	.065%	−.027%	.065%
cotyledons	β	.394%	1.502%	1.502	.394	−4.137%	1.485%
	γ	.415%	.415%	−.157%	−.157%	−.157%	.415%
	PE	.351	.527	.665	.220	.205	.319
Bean root	α	.071%	−.034%	−.034%	.071%	−.034%	.071%
cells	β	.438%	5.346%	5.346%	.438%	1.891%	5.163%
	γ	.444%	.444%	−.177%	−.177%	−.177%	.444%
	PE	.372	1.964	2.266	.347	.345	1.837

skewness coefficient and coefficient of excess (kurtosis), and also tend to have significant bias and a variance in excess of the MVB (see Sec. 2.6 for details of how to carry out these tests). Those parameterizations for which the calculated percentage biases in the LS estimators are less than 1% for each data set are the ones which come closest to exhibiting linear behavior; generally, the distributions of the LS estimates for these parameters, from the results of the simulation studies, are either not significantly different from the normal, or are only slightly different. Unlike the model functions of the Gompertz model considered in Sec. 4.4.1, these conclusions do not alter when the multiplicative error assumption is considered. Model functions such as (4.2) and (4.9) thus appear to possess a sufficiently close approach to linear behavior for practical use.

4.4.3 Richards model

Of the five models studied in this chapter, the Richards model is the only one exhibiting a significant intrinsic curvature. This fact militates against spending too much effort searching for better parameterizations of model function (4.3) initially considered,

$$Y = \frac{\alpha}{[1 + \exp (\beta - \gamma X)]^{1/\delta}}$$

as even the "best" reparameterization, the one whose parameter-effects nonlinearity is zero, will still have the same intrinsic nonlinearity. Nevertheless, for interest and completeness, the following two other model functions of the Richards model are studied:

$$Y = \frac{\alpha}{\{1 + \beta[\exp (-\gamma X)]\}^{1/\delta}} \qquad (4.12)$$

$$Y = \frac{\alpha/\beta}{\{1 + \exp [-\beta(1 + \delta)(X - \gamma)]\}^{1/(1+\delta)}} \qquad (4.13)$$

The last model function has a cumbersome appearance due to the fact that two of its parameters appear twice, but it is of historical interest as it is a generalized form of the equation derived by von

Bertalanffy (1951) from studies of allometric relations in organisms.

Table 4.6 presents results for the curvature measure of parameter-effects nonlinearity of Bates and Watts, and for the bias calculation of Box for the three model functions considered. The results clearly show that model functions (4.12) and (4.13) are significantly worse than (4.3). Of interest is the enormous percentage bias for the estimates of parameter β in (4.12), being of the order

Table 4.6 Richards Model: Box's Percentage Bias and Parameter-Effects (PE) Curvatures for Model Functions (4.3), (4.12), and (4.13): Additive Error Assumption

Data set	Parameter	Model function		
		(4.3)	(4.12)	(4.13)
Pasture	α	.376%	.376%	5.520%
regrowth	β	2.305%	90.125%	2.658%
	γ	2.447%	2.447%	-1.957%
	δ	3.567%	3.567%	8.492%
	PE	6.679	68.693	24.230
Onion bulbs	α	.221%	.221%	17.469%
plus tops	β	5.135%	244.017%	7.583%
	γ	5.250%	5.250%	-3.522%
	δ	8.797%	8.797%	40.297%
	PE	6.271	90.368	37.206
Cucumber	α	.209%	.209%	38.909%
cotyledons	β	1.085%	107.534%	15.559%
	γ	5.108%	5.108%	-17.743%
	δ	11.599%	11.599%	681.391%
	PE	14.811	48.540	93.906
Bean root	α	.184%	.184%	8.181%
cells	β	4.431%	192.766%	3.684%
	γ	4.232%	4.232%	-1.843%
	δ	6.363%	6.363%	16.644%
	PE	4.268	80.619	21.161

of 100 to 200%. A simulation study reveals that the estimators of β in this model have positive skewness with a long right-hand tail, suggestive of a lognormal distribution (see Exercise 4.6). This indicates that it would be beneficial to replace β in (4.12) by exp (β), which of course leads to model function (4.3). Although (4.3) is the best of the three model functions of the Richards model considered here, the percentage biases of β, γ, and δ are much too high for one to consider using this model in practice. Fortunately, there are suitable alternatives to this model, as will be seen in the next two subsections.

4.4.4 Morgan-Mercer-Flodin (MMF) model

The MMF model (4.4),

$$Y = \frac{\beta\gamma + \alpha X^{\delta}}{\gamma + X^{\delta}}$$

was presented by its originators (Morgan et al., 1975) as an extension of two well-known models in use in catalytic kinetic studies. When $\beta = 0$, model function (4.4) reduces to the Hill (1913) model and when $\beta = 0$ and $\delta = 1$, to the familiar Michaelis-Menten (1913) rectangular hyperbola. The parameter β in (4.4) allows the model to have a nonzero intercept on the Y-axis.

Although model function (4.4) possesses a high degree of parameter-effects nonlinearity for each of the four data sets considered here (Table 4.2), it appears from the percentage bias in Table 4.3 that most of this nonlinearity may be contained in parameter γ, suggesting that a suitable reparameterization involving this parameter may markedly reduce the nonlinearity. In order to obtain an idea of which reparameterization might be suitable, a simulation study of model function (4.4) was carried out for each of the four data sets. A histogram of the LS estimates $\hat{\gamma}$ is presented in Fig. 4.2a for pasture regrowth only, but the other three crops give a similar picture of a highly positively skewed distribution with a long right-hand tail. This form is suggestive of a lognormal distribution and indicates that

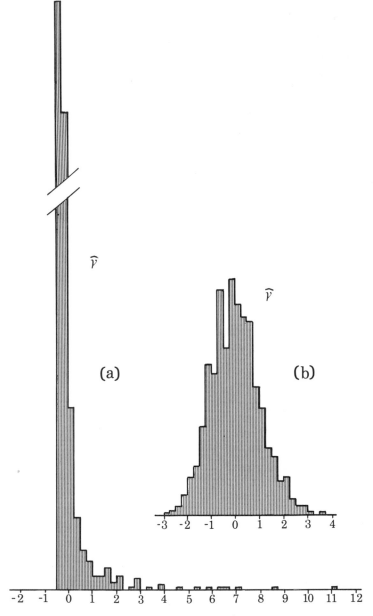

Fig. 4.2 Histogram of results for $\hat{\gamma}$ in (a) original MMF model and (b) reparameterized MMF model; pasture regrowth data set.

model function (4.4) may be improved by replacing γ by its exponential.
Thus the new model function to be considered is

$$Y = \frac{\beta \exp (\gamma) + \alpha X^\delta}{\exp (\gamma) + X^\delta} \qquad (4.14)$$

To make a preliminary assessment as to whether the percentage bias
in the estimate of γ might be substantially reduced in the new model
function (4.14), use may be made of formula (2.7). The procedure will
be illustrated using the pasture regrowth data.

$$\hat{\gamma} \ [\text{in } (4.4)] = 49,577.3$$

$$\text{Var} (\hat{\gamma}) = 5.33858(10^9) \ [\text{from the asymptotic covariance matrix}]$$

$$\text{Bias} (\hat{\gamma}) = 59,231.4 \quad [\% \text{ bias} = \frac{100(59,231.4)}{49,577.3} = 119.5\%]$$

The "new" parameter in (4.14) is related to the "old" γ in (4.4) by

$$g(\gamma) = \log \gamma$$

Therefore,

$$g(\hat{\gamma}) = \log \hat{\gamma} = 10.811$$

$$\frac{\partial g(\gamma)}{\partial \gamma} = \frac{1}{\hat{\gamma}} = 2.0171(10^{-5})$$

$$\frac{\partial^2 g(\gamma)}{\partial \gamma^2} = \frac{-1}{\hat{\gamma}^2} = -4.0685(10^{-10})$$

and so from formula (2.7),

$$\text{Bias} [g(\hat{\gamma})] = \text{Bias}(\hat{\gamma}) \frac{\partial g(\gamma)}{\partial \gamma} + \frac{1}{2} \text{Var} (\hat{\gamma}) \frac{\partial^2 g(\gamma)}{\partial \gamma^2}$$

$$= 59,231.4(2.0171)(10^{-5})$$

$$+ \frac{1}{2}(5.33858)(10^9)(-4.0685)(10^{-10})$$

$$= 1.1947 - 1.0860 = .1087$$

$$\% \text{ Bias} [g(\hat{\gamma})] = \frac{100(.1087)}{10.81} = 1.006\%$$

Thus, the relative bias in parameter γ, which was previously more than
100%, is reduced to only 1% as a result of the new parameterization
(4.14).

Complete results are presented in Table 4.7 for all data sets. In each case, the parameter-effects nonlinearity is considerably reduced as a result of the reparameterization. Although the behavior is still significantly nonlinear in the statistical sense, the extent of the nonlinearity is much less than before. A histogram of the LS estimates $\hat{\gamma}$ from a simulation study of model function (4.14) for pasture regrowth is presented in Fig. 4.2b; these estimates are much closer to a normal distribution than the corresponding LS estimates $\hat{\gamma}$ of γ

Table 4.7 MMF Model: Box's Percentage Bias and Parameter-Effects (PE) Curvatures for Model Functions (4.4) and (4.14): Additive Error Assumption

Data set	Parameter	Model function	
		(4.4)	(4.14)
Pasture	α	1.525%	1.525%
regrowth	β	−1.645%	−1.645%
	γ	119.478%	1.006%
	δ	.921%	.921%
	PE	90.970	4.406
Onion bulbs	α	.313%	.313%
plus tops	β	−2.483%	−2.483%
	γ	86.600%	1.468%
	δ	1.424%	1.424%
	PE	31.319	1.210
Cucumber	α	.040%	.040%
cotyledons	β	−.047%	−.047%
	γ	1.272%	.197%
	δ	.172%	.172%
	PE	1.154	.497
Bean root	α	.237%	.237%
cells	β	−.794%	−.794%
	γ	55.341%	.972%
	δ	.942%	.942%
	PE	24.934	1.067

for model function (4.4) in Fig. 4.2a. Similar results for $\hat{\gamma}$ are
obtained for the other data sets (see Exercise 4.7). In no case is
there any drastic departure of the LS estimates of any parameter from
normality, although in many cases the departures are significant
statistically. The worst departure from normality occurs for $\hat{\alpha}$ for
pasture regrowth, and it is this parameter that is mainly responsible
for the relatively high parameter-effects curvature exhibited by this
data set. The other three data sets have much lower parameter-effects
curvatures and in all cases the departures from normality of the LS
estimates, although sometimes significant statistically, are slight
in practical terms.

4.4.5 Weibull-type model

The Weibull-type model was originally studied in model function (4.5),

$$Y = \alpha - \beta \exp(-\gamma X^{\delta})$$

Although the parameter-effects nonlinearity is high for (4.5) for each
of the four data sets considered here (Table 4.2), it appears from the
percentage bias in Table 4.3 that most of this nonlinearity may be
contained in parameter γ, suggesting that a suitable
reparameterization involving this parameter could markedly reduce the
nonlinearity. A very similar situation occurred with the MMF model.
To obtain some indication of a suitable reparameterization, a
simulation study of model function (4.5) was carried out for each of
the four data sets. A histogram of the LS estimates $\hat{\gamma}$ is presented
in Fig. 4.3a for pasture regrowth; the other three crops also give a
similar picture of a highly positively skewed distribution with a long
right-hand tail. As this form is suggestive of a lognormal
distribution, it appears that model function (4.5) may be improved by
replacing γ by $\exp(\gamma)$. However, as $\hat{\gamma}$ in (4.5) is less than unity for
each of the four data sets, we use $\exp(-\gamma)$ instead of $\exp(\gamma)$, so that
values of $\hat{\gamma}$ in the new parameterization shall be positive. Thus the
new model function to be considered is as follows:

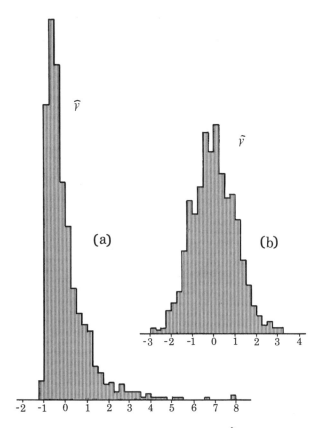

Fig. 4.3 Histogram of results for $\hat{\gamma}$ in (a) original Weibull-type model (4.5) and (b) reparameterized Weibull-type model (4.15); pasture regrowth.

$$Y = \alpha - \beta \exp \left[-\exp \left(-\gamma\right)X^{\delta}\right] \tag{4.15}$$

To make a preliminary assessment as to whether the bias in the estimates of γ in the new model function (4.15) might be substantially less than in (4.5), use is made of formula (2.7) . The procedure exactly parallels that for the MMF model in Sec. 4.4.4.   From formula (2.7), for the pasture regrowth data,

$$\text{Bias} \ [g(\hat{\gamma})] = 2.98895(10^{-5})(-9985.9)$$

$$+ \frac{1}{2} (6.69848)(10^{-9})(9.9719)(10^{7})$$

$$= -.29847 + .33398 = .03551$$

$$\% \text{ Bias } [g(\hat{\gamma})] = \frac{100(.03551)}{9.2089} = .386\%$$

Thus the bias in parameter γ which was previously almost 30% is reduced to less than 1% in the new parameterization (4.15).

Complete results are presented in Table 4.8 for all data sets. In each case, the parameter-effects nonlinearity is considerably reduced as a result of the reparameterization, and although the behavior is still significantly nonlinear in the statistical sense, the extent of the nonlinearity is slight in practical terms. A histogram of the LS estimates $\hat{\gamma}$ from a simulation study of model function (4.15) is presented in Fig. 4.3b for pasture regrowth; these estimates are much closer to a normal distribution than the correponding LS estimates $\hat{\gamma}$ for model function (4.5) in Fig. 4.3a. Similar results for $\hat{\gamma}$ are obtained for the other data sets (see Exercise 4.8). In no case is there any drastic departure of the LS estimates of any parameter from normality, although in many cases the departures are significant statistically. The worst departures from normality are those for $\hat{\alpha}$ and $\hat{\beta}$ for pasture regrowth; the skewness coefficients are positive for the LS estimates of these two paramcters for each of the four data sets. This suggests replacing these parameters by their exponentials in the following reparameterization:

$$Y = \exp (\alpha) - \exp [\beta - \exp (-\gamma)X^{\delta}] \qquad (4.16)$$

Results for the parameter-effects curvature and for the bias calculation of Box are presented in Table 4.8 for the model function above. For each data set the curvature is less than that for model function (4.15) but only by a relatively small amount which is not sufficient to make the nonlinearity not significant statistically. Similarly, the percentage biases in $\hat{\alpha}$ and $\hat{\beta}$ are also reduced for each data set, but the results of simulation studies do not indicate any substantial improvement in normality by the reparameterization given by model function (4.16) (see Exercise 4.8). For example, for pasture

Table 4.8   Weibull-Type Model: Box's Percentage Bias and Parameter-Effects (PE) Curvatures for Model Functions (4.5), (4.15), and (4.16): Additive Error Assumption

Data set	Parameter	Model function		
		(4.5)	(4.15)	(4.16)
Pasture	α	.490%	.490%	.102%
regrowth	β	.686%	.686%	.134%
	γ	29.847%	.386%	.386%
	δ	.333%	.333%	.333%
	PE	42.675	1.631	1.437
Onion bulbs	α	.104%	.104%	.013%
plus tops	β	.186%	.186%	.019%
	γ	16.379%	.917%	.917%
	δ	.916%	.916%	.916%
	PE	16.371	.589	.549
Cucumber	α	.112%	.112%	.050%
cotyledons	β	.156%	.156%	.050%
	γ	.348%	.827%	.827%
	δ	.757%	.757%	.757%
	PE	1.878	.574	.541
Bean root	α	.103%	.103%	.028%
cells	β	.157%	.157%	.036%
	γ	11.485%	.703%	.703%
	δ	.698%	.698%	.698%
	PE	13.253	.540	.510

regrowth, the replacement of α and β by exp (α) and exp (β), respectively, only results in a slight improvement in the normality of the LS estimates; for the other three sets of data, the estimators of α and β are very close to behaving normally initially and the transformation has little effect.   It thus appears that the less complicated model function (4.15) might be adequate for most purposes, as the bias in the parameter estimates is less than 1% for each parameter and the departure from normality is never drastic.

4.5 Choice of Growth Model or Model Function

Section 4.3 was devoted to an examination of the nonlinearity in the five sigmoidal growth models, and Sec. 4.4 was concerned with reparameterizing these models, the object being to find a parameterization in which the parameters come as close as possible to behaving linearly, so that their LS estimators are close to being normally distributed. We now consider each model in turn and summarize what has been learned from these preceding two sections.

The Gompertz model has an acceptable intrinsic curvature of its solution locus, but is the worst fitting of the five models with respect to each of the data sets. If these data are typical, this suggests that this model may not be as suitable as the other models for modeling vegetative growth processes. The other three-parameter model, the logistic model, proves to be a very suitable model for these data sets. It consistently fits the data well, as judged by its relatively low residual variance, and its solution locus consistently has an acceptably low intrinsic curvature. Various model functions of the logistic model were considered; model functions (4.2) and (4.9) consistently have low parameter-effects nonlinearities with biases in the parameter estimates always being less than 1% for each parameter. Simulation studies confirm that the behavior of model functions (4.2) and (4.9) is close to linear both for additive and multiplicative errors. Thus either of these model functions appears to be suitable for modeling vegetative growth processes.

Situations will arise in practice, however, where a three-parameter model such as the logistic model will not provide an adequate fit to certain classes of data and the greater flexibility of a four-parameter model will be necessary. The four-parameter models considered in this chapter are the Richards, MMF, and Weibull-type models. The Richards model may be viewed as a simple extension of the logistic model. In view of the fact that (4.2) is a model function whose behavior closely approaches linear behavior, it is of interest that (4.3) is relatively poorly behaved. The Richards model is the only model that has an unacceptable intrinsic nonlinearity, as the solution locus departs significantly from a hyperplane for each of the four data sets. This alone may be sufficient to induce a modeler to abandon it from further consideration. In addition, Davies and Ku (1977) found that for a somewhat different model function of the Richards model from the ones considered here, three of its four parameters were poorly estimated by least squares, there being a very wide range of values for each of these parameters, covering several orders of magnitude, which gave almost identically the same minimum sum of squares. This provides further evidence (see Secs. 4.2 and 4.3) that instability in parameter estimates is closely related to nonlinear behavior of the model.

The MMF model function (4.14) exhibited reasonably close-to-linear behavior. However, for the pasture regrowth data set, the distribution of $\hat{\alpha}$ departed substantially from normality, in contrast to the cucumber cotyledon data set, where the parameter-effects curvature was only a little in excess of the requirement for acceptable linearity, and where the percentage biases were less than 0.2% for each of the four parameters. The MMF model fitted the cucumber data extremely well, the residual variance being considerably smaller than for the other four models. As the bias measure of Box is directly proportional to the residual variance [see formula (2.3) for verification of this], and as the curvature measures of nonlinearity of Bates and Watts are proportional to the square root of the residual variance, the low values of these measures for the cucumber data compared with the other data sets are readily explained. The good fit of the MMF model to the cucumber data, consisting of nine observations, is possibly due to chance, and may not indicate that this model is better suited to such data than alternative models. The behavior of model function (4.14) with the remaining two data sets, onion bulbs and tops and bean root cells, was reasonably satisfactory, as in no case did the estimators of any parameter depart more than slightly from linear behavior.

The Weibull-type model function (4.15) behaved very satisfactorily, its worst behavior occurring with the pasture regrowth data set, where $\hat{\alpha}$ and $\hat{\beta}$ departed significantly from linear behavior, although not drastically so. For each of the other three data sets, the parameter-effects nonlinearity was not far in excess of that required for acceptable linearity, and in all cases the percentage bias in each of the estimates was less than 1%. Another parameterization was considered, namely model function (4.16), but this resulted in only a relatively small gain in linear behavior compared with model function (4.15).

Another potentially useful criterion for examining the acceptability of a model function is Student's t (see Sec. 2.8). Table 4.9 presents t-values for the parameter estimates for each of the five models studied in this chapter, using their best or nearly best model functions. These values are very high for the model functions of the Gompertz, logistic, and Weibull-type models, indicating that the parameter estimates are well determined in these model functions. For the Richards model, the t-values associated with the estimates of β, γ, and δ in (4.3) are relatively low and in several cases are not significant. This behavior closely parallels the results in Table 4.6 for this model function, where the estimates of these same parameters were the ones with high percentage biases. For the MMF model, the t-values associated with the estimates of all parameters of (4.14) are generally high, although the value corresponding to parameter β for onion bulbs and tops is not significant; it was also the estimate of this same parameter for this same data set which had the highest percentage bias for this model function (see Table 4.7). To demonstrate further the close connection between t-values and possible nonlinear behavior of parameter estimators, t-values are presented in Table 4.10 for parameter γ in the MMF and Weibull-type models in the original parameterization [i.e., model functions (4.4) and (4.5), respectively] and also after reparameterization [model functions (4.14) and (4.15), respectively]. It is clear how much higher the t-values are when a parameter appears in a more suitable parameterization. In fact, except

Table 4.9 Student's t-Values, Being the Ratios of the Parameter
Estimates to Their Standard Errors

Data set	Parameter	Gompertz (4.1)	Logistic (4.2)	Richards (4.3)	MMF (4.14)	Weibull (4.15)
Pasture	α	14.54	41.79	32.76	12.15	29.62
	β	16.28	29.65	3.36	5.35	19.32
	γ	8.52	19.54	4.83	7.34	11.27
	δ			2.91	6.65	10.76
Onion	α	32.78	50.42	42.92	30.12	52.79
	β	8.78	12.66	2.53	1.69	28.09
	γ	8.69	12.00	3.88	7.20	9.71
	δ			1.86	7.15	9.75
Cucumber	α	31.71	50.78	38.85	80.76	52.93
	β	8.32	13.82	1.22	19.07	26.54
	γ	9.56	13.52	3.89	20.67	10.39
	δ			1.40	20.68	10.90
Bean	α	26.88	51.78	47.80	34.42	55.69
	β	8.96	15.11	3.11	3.80	31.96
	γ	8.46	13.95	4.36	8.90	11.22
	δ			2.43	8.73	11.17

Table 4.10 Student's t-Values for the Estimates of Parameter γ in MMF
and Weibull-Type Models Before and After Reparameterization

	MMF model function		Weibull-type model function	
Data set	(4.4)	(4.14)	(4.5)	(4.15)
Pasture regrowth	.68	7.34	1.22	11.27
Onion bulbs plus tops	.82	7.20	1.50	9.71
Cucumber cotyledons	8.07	20.67	4.86	10.39
Bean root cells	1.03	8.90	1.77	11.22

for the cucumber cotyledon data set, the t-values are not significant statistically in model functions (4.4) and (4.5).

Although it is wise to avoid making sweeping generalizations about the suitability or unsuitability of the various models or model functions studied in this chapter, it is tempting to recommend the logistic model in model function (4.2) as a suitable three-parameter model for vegetative growth modeling. There may, of course, be data sets for which this model function does not perform well. Nevertheless, in view of the information obtained about its performance on the four data sets studied in this chapter, model function (4.2) should be the first choice for consideration. Similarly, when a four-parameter model is required for modeling vegetative growth processes, the Weibull-type model in model function (4.15) and the MMF model in model function (4.14) appear to be suitable, whereas the Richards model in any of the model functions studied here is rather unsuitable.

4.6 Interpretation of the Parameters in the Sigmoidal Growth Models

Throughout this chapter, the same Greek letters, α, β, γ, and δ, have been used to denote the parameters, not only in the original model functions (4.1)-(4.5), but also in the various reparameterizations (4.6)-(4.16). The use of the same notation was deliberate, as the parameters indicated by the same Greek letter have a consistent meaning. Consider parameter α, which in all model functions is the parameter relating to the asymptote. For most of the model functions, the asymptote is just α, but in (4.6) it is exp (α), and in (4.7), (4.8), and (4.10) it is the reciprocal of α. The parameter β relates to the "intercept" on the Y-axis (i.e., the Y-value corresponding to X = 0). In some model functions, such as (4.4) and (4.14), the intercept is just β; in others, such as (4.5), it is the difference between the asymptote and β. In all cases, however, this parameter determines the position of the intercept. The parameter γ relates to the rate at which the response changes from its "initial" value (determined by the magnitude of β) to its "final" value (determined by the magnitude of α). The parameter δ is present in the four-parameter models to provide increased flexibility to the models for data fitting, in comparison with the three-parameter models.

It was stressed in Sec. 2.8 that barriers to understanding should not arise simply as a result of a monotonic transformation of a parameter. For example, if α relates to the asymptote, as it does for the models in this chapter, then exp (α), log α, and 1/α also relate to the asymptote. In comparing results obtained from different data sets, the set with the highest α has the highest asymptote for any of the formulations above, except 1/α, in which case the data set with the lowest α has the highest asymptote. As long as monotonicity of transformation is maintained, problems of user resistance to reparameterized models should not be encountered.

Appendix 4.A: Data Sets

Data set 1[a]		Data set 2[b]		Data set 3[c]		Data set 4[d]	
X	Y	X	Y	X	Y	X	Y
9	8.93	1	16.08	0	1.23	.5	1.3
14	10.80	2	33.83	1	1.52	1.5	1.3
21	18.59	3	65.80	2	2.95	2.5	1.9
28	22.33	4	97.20	3	4.34	3.5	3.4
42	39.35	5	191.55	4	5.26	4.5	5.3
57	56.11	6	326.20	5	5.84	5.5	7.1
63	61.73	7	386.87	6	6.21	6.5	10.6
70	64.62	8	520.53	8	6.50	7.5	16.0
79	67.08·	9	590.03	10	6.83	8.5	16.4
		10	651.92			9.5	18.3
		11	724.93			10.5	20.9
		12	699.56			11.5	20.5
		13	689.96			12.5	21.3
		14	637.56			13.5	21.2
		15	717.41			14.5	20.9

Exercises

4.1 Carry out simulation studies of 1000 trials for the logistic model in model function (4.2) assuming additive error, for each of the four data sets of Appendix 4.A, and show that the distribution of the set of estimates of each parameter is close to that of a normal distribution. Repeat the simulation studies using multiplicative error. Use the parameter estimates and residual variances given in Table 4.1.

4.2 Derive formulas relating the LS estimators of the parameters of model function (4.6) of the Gompertz model with the parameters of model function (4.1) of the same model. Determine thereby the LS estimates of the parameters of model function (4.6) from the LS estimates of model function (4.1) tabulated in Table 4.1, assuming additive error for each of the four data sets.

4.3 Carry out simulation studies of 1000 trials for the Gompertz model in model function (4.6) assuming additive error, for each of the four data sets of Appendix 4.A, and show that the estimates of parameter β deviate markedly from a normal distribution.

4.4 Calculate Box's bias measure for model function (4.1), assuming multiplicative error, for each of the four data sets of Appendix 4.A, and show that the percentage bias in the estimator of α is greater than 1% for each data set. Carry out simulation studies of 1000 trials each for each data set, and show that the LS

---

[a]Pasture yield (Y) versus growing time (X). (Courtesy of W. F. Hunt, unpublished data.)

[b]Onion bulbs plus tops dry weight (Y) versus growing time (X). (Courtesy of B. M. Beattie, unpublished data.)

[c]Area of cucumber cotyledons (Y) versus growing time (X). (Data from Gregory, 1956.)

[d]Water content of bean root cells (Y) versus distance from tip (X). (Data from Heyes and Brown, 1956.)

estimators of parameter α are significantly biased and non-normal in distribution.

4.5   Carry out simulation studies of 1000 trials for the logistic model in model functions (4.7), (4.9), and (4.10) assuming additive error, for each of the four data sets of Appendix 4.A, and show that the LS estimators whose calculated percentage biases in Table 4.5 exceed 1% all have non-normal distributions, whereas those estimators with calculated biases less than 1% are close to normal in distribution.

4.6   Carry out simulation studies of 1000 trials for the Richards model in model function (4.3) assuming additive error, for each of the four data sets of Appendix 4.A. To ensure convergence to the LS estimates of the parameters for each of the 1000 trials, good initial parameter estimates are required and it will be necessary to use the methods described in Sec. 8.3.3. Study the behavior of the LS estimators of model function (4.12) and (4.13) by saving the estimates from (4.3) and suitably transforming them. For example, exponentiating the LS estimate of β in (4.3) gives the LS estimate of β in (4.12).

4.7   Carry out simulation studies of 1000 trials for the reparameterized MMF model in model function (4.14) for each of the four data sets of Appendix 4.A. Carry out the tests detailed in Sec. 2.6 to examine how closely the set of estimates for each parameter approaches that expected of a linear model. Save the estimates of parameter γ and transform them by taking their exponential. This gives a set of estimates of γ in the original parameterization, model function (4.4). Show, by using the tests of Sec. 2.6, that this results in highly biased, positively skewed, non-normal distributions. Also draw histograms of your set of estimates and see whether they are similar to those depicted in Fig. 4.2.

4.8 Carry out simulation studies of 1000 trials for the reparameterized Weibull-type model in model function (4.15) for each of the four data sets of Appendix 4.A. Carry out the tests detailed in Sec. 2.6 to examine how closely the set of estimates for each parameter approaches that expected of a linear model. Save the estimates of parameters α, β, and γ and suitably transform them so as to be able to study the estimation properties of γ in the original parameterization, model function (4.5), and of α and β in the reparameterization given by model function (4.16). Use the tests of Sec. 2.6 and see whether your conclusions are the same as those reached in Section 4.4.5.

5

Asymptotic Regression Model

5.1 Introduction

Asymptotic regression models with the deterministic component of the form

$$Y = \alpha - \beta\gamma^X \tag{5.1}$$

have been used extensively in agriculture, and to a lesser extent in biology and the engineering sciences. One of the earliest and most frequent applications of this model has been in fertilizer experimentation, where X is the rate of fertilizer application and Y is the yield of the crop. In this case the model is often referred to as Mitscherlich's law. The model is often used in fisheries research to relate the length of a fish with its age, and there it is often referred to as the von Bertalanffy growth curve (see, e.g., Radway Allen, 1966, or Phillips and Campbell, 1968). The model is similar to the growth models studied in Chapter 4 in that the curve approaches an asymptote with increasing X, but differs in that is lacks an inflection point and hence is not sigmoidal in shape.

Seven data sets, which come from a variety of sources, will be examined in this chapter. The raw data are listed in Appendix 5.A and concern the following responses: (1) length versus age for dugongs; (2) leaf production versus light irradiance; (3) wheat yields versus rate of fertilizer application; (4) chemical reaction versus time; (5) wheat yields versus rate of fertilizer application; (6) potato yields

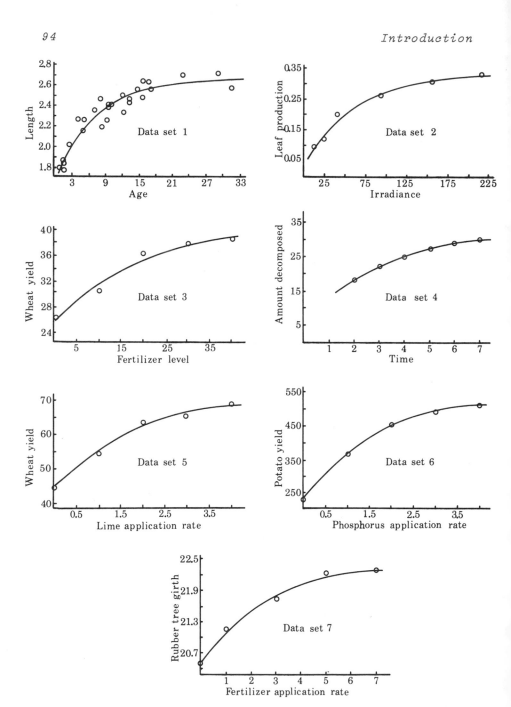

Fig. 5.1 Graph of data (Appendix 5.A) for asymptotic regression model.

versus rate of fertilizer application; and (7) rubber tree girth versus rate of fertilizer application. The data are shown graphically in Fig. 5.1 and it appears that the assumption of an "additive" stochastic term is appropriate.

Whereas attention was directed initially in Chapter 4 to making a comparison between various competing models, namely (4.1)-(4.5), the present chapter concerns only a single model, the asymptotic regression model. In addition to model function (5.1), however, we will examine various other model functions of that model in an effort to find one in which the behavior of the LS estimators approximates linear behavior as closely as possible. Of the infinite number of possible reparameterizations of (5.1), several are considered in the present work. Results will be presented for the following model functions:

$$Y = \alpha - \beta \exp (-\gamma X) \qquad (5.2)$$
$$Y = \alpha\{1 - \exp [-(X + \beta)\gamma]\} \qquad (5.3)$$
$$Y = \alpha - \exp [-(\beta + \gamma X)] \qquad (5.4)$$
$$Y = \alpha - \exp (-\beta)\gamma^X \qquad (5.5)$$
$$Y = \frac{1}{\alpha} - \beta\gamma^X \qquad (5.6)$$
$$Y = \exp (\alpha) - \beta\gamma^X \qquad (5.7)$$

Model functions (5.2)-(5.4) have been considered because they are frequently used in practice.

5.2 Examination of Nonlinearity in the Asymptotic Regression Model

The LS estimates of the parameters of model function (5.1) and the residual variance $\hat{\sigma}^2$ are given in Table 5.1 for each of the seven data sets assuming additive error. For each data set and each model function, Table 5.2 gives the intrinsic and parameter-effects curvatures, which are assessed by comparing them to the critical values $1/(2\sqrt{F})$, as detailed in Sec. 2.4, for significance level $\alpha = 0.05$.

Data sets 1, 4, and 6 have acceptably low intrinsic curvatures (Table 5.2), whereas the solution loci for the other data sets are significantly curved. Since the residual degrees of freedom ν is only

Table 5.1   LS Estimates of Parameters α, β, and γ of Model Function
(5.1) and Residual Variance $\hat{\sigma}^2$, Assuming Additive Error

Data set	$\hat{\alpha}$	$\hat{\beta}$	$\hat{\gamma}$	$\hat{\sigma}^2$
1. Dugong length	2.66663	.972536	.873499	$.778163(10^{-2})$
2. Leaf production	.335021	.295512	.983830	$.201878(10^{-3})$
3. Wheat yield	41.7079	15.8388	.956284	1.45447
4. Chemical reaction	33.8023	26.6980	.752994	$.349154(10^{-1})$
5. Wheat yield	72.4326	28.2519	.596790	1.78440
6. Potato yield	539.078	307.542	.537460	28.0625
7. Rubber tree girth	22.4870	1.95860	.705539	$.587643(10^{-2})$

Table 5.2   Intrinsic and Parameter-Effects Curvatures for Original
X-Values

Data set	Intrinsic curvature	Parameter-effects curvature for model function							Critical value $1/(2\sqrt{F})$
		(5.1)	(5.2)	(5.3)	(5.4)	(5.5)	(5.6)	(5.7)	
1	.207	1.41	1.53	4.18	1.54	1.43	1.22	1.31	.288
2	.232	1.53	1.54	2.56	1.54	1.53	1.21	1.36	.164
3	.246	6.98	7.14	16.28	7.79	7.64	5.37	6.15	.114
4	.042	1.80	2.31	4.24	2.32	1.81	1.09	1.42	.164
5	.148	2.43	3.21	7.09	3.41	2.64	1.82	2.11	.114
6	.053	.57	.80	1.32	.83	.60	.40	.48	.114
7	.150	1.62	1.94	10.81	2.02	1.70	1.55	1.59	.114

either two or three for all data sets except data set 1, the significant
curvature is not surprising (see Exercise 5.1). On the other hand,
the parameter-effects curvatures are unacceptable for all model
functions for all data sets, the worst-behaved model function being
(5.3). The high parameter-effects curvatures for model function (5.3)
are also reflected in high percentage biases in the estimates of its
parameter β; it appears that that model function would be a poor choice
for practical use.

Converting parameter γ from the form γ^X to the form $\exp(-\gamma X)$
increases the nonlinearity [compare the curvatures for model functions

(5.2) and (5.1)]. Similarly, converting parameter β to the form exp (−β) generally results in an increase in nonlinearity [(5.5) versus (5.1)]. Model function (5.4) is a combination of the latter two transformations. These curvatures are almost additive, in the sense that the sum of the curvatures for model functions (5.1) and (5.4) is almost equal to the sum of the curvatures for model functions (5.2) and (5.5). Converting parameter α to the form exp (α) results in an improvement in the linearity [(5.7) versus (5.1)]. Converting α to its reciprocal 1/α improves the linearity even more, and thus (5.6) appears to be the best of the model functions examined in this chapter.

5.3 Positioning the Design Values X

We now examine the question of how the design values of the regressor variable X may influence the linear behavior of the asymptotic regression model. Box and Lucas (1959) studied this question for nonlinear models in general by assuming that the asymptotic properties of the LS estimators adequately approximated the true situation, and that the minimization of Wilks's generalized variance was a suitable criterion for optimality. By confining themselves to the case where the number of design points equaled the number of parameters to be estimated, they were able to show for the asymptotic regression model that the optimum design involves choosing one of the values of X to be as near zero as possible, another to be sufficiently large to give an almost maximal response, and the third to be such as to yield a response equal to 63.2% of the maximum. For more than three design points, the determination of the optimum conditions is much more complicated and involves tedious algebraic manipulation. Box and Lucas (1959) remarked that each of the three basic trial points can be replicated r times, to obtain any desired degree of precision in the LS estimates. However, that arrangement does not necessarily minimize the generalized variance for all possible experiments involving 3r observations, and the parameter-effects curvature may be far from optimal. Because of these considerations, in this chapter we adopt the procedure of trying to obtain a relatively even spread of X-values throughout their range, making sure that the lowest value is close to zero and that the highest value is such as to give an almost maximal response.

The effect of changing the values of the design variables was studied for each of the seven data sets of this chapter. The oldest dugong in data set 1 was only 31.5, although Marsh (1980) gives the life span of a dugong as between 50 and 60 years. If a dugong of age 60 years had been available, and if the estimates given in Table 5.1 may be assumed to be the "true" parameter values, then the predicted length 2.66634 for age 60 would be within .01% of the maximum length attainable. To provide a relatively even spread of ages through the range one to 60, the following hypothetical values were used: 1, 2, 4, 6, 8, 10, 12, 14, 16, 18, 20, 22.5, 25, 27.5, 30, 32.5, 35, 37.5, 40, 42.5, 45, 47.5, 50, 52.5, 55, 57.5, 60. The sample size of 27 data points was thus the same as in the original, and the parameter estimates and residual variance were also taken to be the same; the only change involved the values of the regressor variable X. A similar approach was taken with the other data sets, for example for data set 2, a maximum value of 500 was used, instead of 215 in the actual data, and the remaining design points were spread about through the range, while maintaining the total sample size at the actual original number of six.

Results for the recalculated parameter-effects curvatures are presented in Table 5.3 for model functions (5.1) and (5.6). A considerable reduction in nonlinearity is achieved in practically all data sets, although in every case the parameter-effects curvature is significant statistically. The difference in curvature between model functions (5.1) and (5.6) is small, and is much less than it was with the original disposition of the X-values (see Table 5.2).

We now look at the values of Box's bias measure (Sec. 2.5) to help identify which parameters contribute most to the parameter-effects nonlinearity. Results in Table 5.4 for model function (5.1) show that when the original set of X-values is used, considerable positive bias may be present in the estimates of parameters α and β; these parameters

Table 5.3   Intrinsic and Parameter-Effects Curvatures for Altered X-Values

Data set	Intrinsic curvature	Parameter-effects curvature for model function		Critical value $1/(2\sqrt{F})$
		(5.1)	(5.6)	
1	.243	.545	.528	.288
2	.204	.523	.475	.164
3	.228	.966	.938	.114
4	.045	.180	.177	.164
5	.178	1.098	1.071	.114
6	.063	.428	.422	.114
7	.141	.660	.656	.114

Table 5.4   Percentage Bias in Parameter Estimates for Model Function (5.1)

Data set	Original X-values			Altered X-values		
	α	β	γ	α	β	γ
1	.233	1.042	−.163	.057	.684	−.150
2	.887	1.603	−.022	.132	.543	−.020
3	3.941	10.481	−.114	.158	.433	−.184
4	.231	.460	.020	.006	.108	−.029
5	1.013	2.633	.337	.106	.274	−1.976
6	.117	.208	.036	.010	.018	−.559
7	.108	1.295	−.032	.011	.133	−.590

may be overestimated by about 4 to 10%, respectively. With the altered set of X-values, both these parameters are estimated to within 1%, although the bias in the estimation of parameter γ has increased somewhat in some data sets.

Simulation studies (Sec. 2.6) for each of the data sets show a close correspondence between the percentage bias reported in Table 5.4

for the original set of X-values and the nonlinear behavior of the parameter estimators. For example, for parameters α and β, data sets 3 and 5 show the worst departures from linear behavior. On the other hand, the LS estimators for parameter γ are generally well behaved, exhibiting reasonably close-to-linear behavior for all data sets. Some further simulation studies were carried out as described in Sec. 2.9 to estimate the asymmetry measure of bias for four of the data sets in combination with model function (5.1). For three of the data sets, the use of (2.16) and (2.18) indicated substantial nonlinear behavior in the estimators of parameters α and β, but only slight nonlinear behavior in the estimator of γ. For data set 6, the nonlinearity was slight for the estimators of all three parameters. The estimator of α is positively biased for each data set, so the asymptote is consistently overestimated. When the positions of the X-values are altered so that they cover more fully the range of the response, this overestimation of the asymptote is not significant.

5.4   Choice of Model Function for the Asymptotic Regression Model

All model functions considered in this chapter, namely (5.1)-(5.7), have unacceptably high parameter-effects curvatures; the estimator of at least one parameter in each model function exhibits significant nonlinear behavior. A considerable reduction in the nonlinear behavior of the LS estimators was achieved by spacing the X-values more uniformly such that Y ranges from its minimum value (at X = 0) to close to its maximum value (α or some monotonic function of α). This suggests that the choice of values of the regressor variable for the asymptotic regression model may be far more important than the choice of parameterization. An optimum choice of a set of X-values that minimizes both the parameter-effects nonlinearity and the bias in the LS estimates for each parameter separately is difficult or perhaps impossible to achieve. In practical terms it may be sufficient to choose X-values that span the range of the response variable with at least some responses being close to the maximum attainable response.

Under this circumstance, model function (5.1) appears to be reasonably adequate for practical purposes, as the bias in the parameter estimates can usually be kept to less than 1% for each parameter.

Model function (4.5) of the Weibull-type model considered in Chapter 4,

$$Y = \alpha - \beta \exp (-\gamma X^{\delta})$$

is an extension of the asymptotic regression model; (4.5) is simply model function (5.2) with an extra parameter δ. This parameter is responsible for the flexibility of the Weibull-type model, enabling it to fit sigmoidal curves, whereas if $\delta = 1$ the model cannot fit data possessing an inflection point. It is of interest that it is not possible to predict the behavior of a model function from the behavior of closely related models. It has already been remarked in Sec. 4.5 that the generally well-behaved properties of the logistic model do not carry over to the Richards model, even though the latter model is simply an extension of the logistic through the introduction of an extra parameter δ. Similarly, parameter γ in model (4.5) gives rise to a badly behaved LS estimator and required reparameterization (Sec. 4.4.5), whereas the estimator of the same parameter in the closely related model function (5.2) is generally well behaved.

Appendix 5.A: Data Sets

Data Set 1: Length (Y) versus age (X) for the sirenian species *Dugong dugon* Müller captured near Townsville, Queensland. The method of determining age has subsequently been revised since this data set was obtained, and the values given below are preliminary data only. (Data courtesy of Dr. Helene Marsh, James Cook University, North Queensland.)

X	1	1.5	1.5	1.5	2.5	4.0	5.0	5.0	7.0
Y	1.80	1.85	1.87	1.77	2.02	2.27	2.15	2.26	2.35

X	8.0	8.5	9.0	9.5	9.5	10.0	12.0	12.0	13.0
Y	2.47	2.19	2.26	2.40	2.39	2.41	2.50	2.32	2.43

X	13.0	14.5	15.5	15.5	16.5	17.0	22.5	29.0	31.5
Y	2.47	2.56	2.65	2.47	2.64	2.56	2.70	2.72	2.57

Data Set 2: Leaf production, number of leaves per tiller per day (Y) versus light irradiance (X) in watts per square meter at 20°C. (Data courtesy of Dr. W. F. Hunt, DSIR Grasslands Division, Palmerston North, New Zealand.)

X	12	23	40	92	156	215
Y	.094	.119	.199	.260	.309	.331

Data Set 3: Wheat yields (Y) versus fertilizer levels (X). [Reprinted by permission from Statistical Methods, 6th ed., by G. W. Snedecor and W. G. Cochran. (C) 1967 by the Iowa State University Press, Ames, Iowa 50010.]

X	0	10	20	30	40
Y	26.2	30.4	36.3	37.8	38.6

Data Set 4: Chemical reaction, amount of N_2O_5 decomposed (Y) versus time (X). [Reprinted by permission from Statistical Methods, 6th ed., by G. W. Snedecor and W. G. Cochran. (C) 1967 by the Iowa State University Press, Ames, Iowa 50010].

X	2	3	4	5	6	7
Y	18.6	22.6	25.1	27.2	29.1	30.1

Data Set 5: Wheat yields (Y) versus rate of lime application (X). [Data from W. L. Stevens, Asymptotic Regression, Biometrics 7, 247-267 (1951). With permission from the Biometric Society.]

X	0	1	2	3	4
Y	44.4	54.6	63.8	65.7	68.9

Data Set 6: Potato yields (Y) versus rate of P_2O_2 application (X). [Data from F. Pimentel-Gomes, The Use of Mitscherlich's Regression Law in the Analysis of Experiments with Fertilizers, Biometrics 9, 498-516 (1953). With permission from the Biometric Society.]

X	0	1	2	3	4
Y	232.65	369.08	455.63	491.45	511.50

Data Set 7: Rubber tree girth (Y) versus rate of fertilizer application (X). (Data from Keeping, 1951.)

X	0	1	3	5	7
Y	20.518	21.138	21.734	22.218	22.286

Exercises

5.1 In Table 5.2, four of the seven data sets have significant intrinsic curvatures. Show that the significant curvature is largely the consequence of low critical values $1/(2\sqrt{F})$ resulting from small sample sizes, and that had the minimum sample size of any data set been n = 12, all intrinsic curvatures would have been acceptable at α = .05.

5.2 Calculate Box's bias measure for model function (5.3) for each of the seven data sets and verify that the estimates of parameter β in this model function tend to have a very high percentage bias. Use the residual variances and parameter estimates from Table 5.1, the latter requiring appropriate transformation since they apply to model function (5.1).

5.3 Having calculated Box's bias measure for one model function, values for other model functions can be calculated using expression (2.3). To demonstrate this, use the results from Exercise 5.2 to calculate the bias in the LS estimates of γ in model (5.1) and, after converting to percentage bias, compare your results with the results in Table 5.4 reported for the "original X values."

5.4 Table 5.3 presents some results for the curvature measures of nonlinearity of Bates and Watts for model functions (5.1) and (5.6) for each of the seven data sets after altering the positions of X. Carry out similar calculations for model functions (5.4) and (5.7). Also examine the values of Box's bias measure expressed as a percentage of the LS estimates for these model functions and compare their magnitudes with the results in Table 5.4 for model function (5.1). The X values should be chosen to span the range of the response variable Y as widely as possible. Use the parameter estimates (after suitable transformation) and residual variances from Table 5.1.

5.5 Carry out simulation studies of 1000 trials for the model

functions considered in Exercise 5.4, with the X-values altered
as suggested there, and examine the connection between the
nonlinear behavior of the LS estimator for each parameter as
determined from the simulation results using the methods of Sec.
2.6, and the percentage bias in the estimate as calculated using
Box's bias measure.

6

Some Miscellaneous Models

6.1 Introduction

Chapters 3–5 studied yield–density models, sigmoidal growth models, and nonsigmoidal growth models, respectively. We now examine a number of other models which are not readily grouped under a single subject heading. We will also examine the effect of sample size on the behavior of the LS estimators of the parameters of a nonlinear model. It will be seen that there are circumstances when the curvature measure of parameter-effects nonlinearity of Bates and Watts predicts that there is significant nonlinearity when in fact the behavior of the LS estimator of each parameter may be close to linear. This illustrates the importance of using several of the techniques for assessing nonlinear behavior together. The combination of (1) the curvature measures of nonlinearity of Bates and Watts, (2) the bias calculation of Box, and (3) simulation studies, is a powerful tool for identifying and assessing the extent of the nonlinear behavior, and for suggesting suitable reparameterizations. The aim of this chapter is again that of finding a model function of each model such that the LS estimators of its parameters come as close as possible to exhibiting linear behavior.

6.2 Model Relating the Age of Wild Rabbits to the Weight of Their Eye Lenses

Dudzinski and Mykytowycz (1961) showed that the weight of the dried eye lens of the European rabbit *Oryctolagus cuniculus* (L.), an animal distributed widely in wild populations in Australia, was a useful indicator of the age of the rabbit. The deterministic component of the model relating eye-lens weight Y to age X may be expressed by

$$Y = \alpha \exp\left(\frac{-\beta}{X + \gamma} \right) \tag{6.1}$$

where α, β, and γ are the parameters of the model. From the graphs in Fig. 6.1 of the set of 71 data observations given in Table 6.1, the variability of the response appears to increase with increasing age, suggesting that a multiplicative error term may be appropriate. Hence the following model function is considered:

$$E(\log Y) = \alpha - \frac{\beta}{X + \gamma} \tag{6.2}$$

The LS estimates of the parameters of (6.2) for the 71 data observations are given in Table 6.2 together with their associated t-values. The relatively high t-values tell us that three parameters of the model are well determined when all 71 observations are used. Results of using the curvature measures of nonlinearity of Bates and Watts (Sec. 2.4) and the bias calculation of Box (Sec. 2.5) are given in Table 6.3.

Both the intrinsic and parameter-effects nonlinearities are sufficiently low for the estimation behavior of the model function/data set combination to be deemed to be close to linear. Note also that the biases in the estimates of the parameters are very small, each being much less than 1% of the estimate.

The effect of sample size on the linear behavior of model function (6.2) is now examined by reducing the 71 data observations in Table 6.1 to 18 observations, and then to 9 observations. To ensure that data are obtained throughout the range of X and Y, a systematic selection procedure using a random starting point is adopted. The original 71 data observations are listed in Table 6.1 in increasing numerical order of X. Using a table of random numbers, and restricting consideration to the numbers one to four, observation 2 (i.e., X_1 = 15, Y_1 = 22.75) was selected. Starting from this observation, retaining every fourth observation produces a set of 18 observations, the final

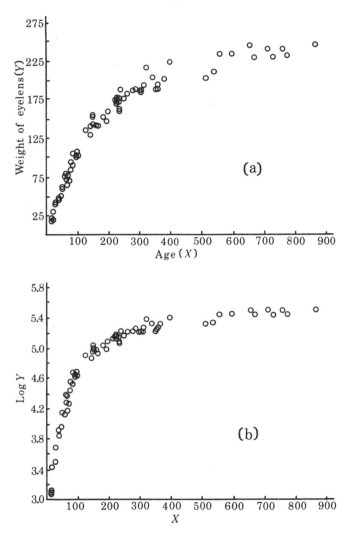

Fig. 6.1 Graph of data (Table 6.1) relating weight of eye lens to age of rabbit: (a) Y versus X; (b) log Y versus X.

one being $(X_{18} = 768, Y_{18} = 232.12)$. This set of 18 observations is further reduced to a set of 9 observations by starting from either the first or second observation and retaining every alternate observation.

Table 6.1 Dry Weight of Eye Lens (Y) in Milligrams as a Function of Age (X) in Days for the European Rabbit in Australia

X	Y	X	Y	X	Y
15	21.66	98	104.30	285	189.66
15	22.75	125	134.90	300	186.09
15	22.30	142	130.68	301	186.70
18	31.25	142	140.58	305	186.80
28	44.79	147	155.30	312	195.10
29	40.55	147	152.20	317	216.41
37	50.25	150	144.50	338	203.23
37	46.88	159	142.15	347	188.38
44	52.03	165	139.81	354	189.70
50	63.47	183	153.22	357	195.31
50	61.13	192	145.72	375	202.63
60	81.00	195	161.10	394	224.82
61	73.09	218	174.18	513	203.30
64	79.09	218	173.03	535	209.70
65	79.51	219	173.54	554	233.90
65	65.31	224	178.86	591	234.70
72	71.90	225	177.68	648	244.30
75	86.10	227	173.73	660	231.00
75	94.60	232	159.98	705	242.40
82	92.50	232	161.29	723	230.77
85	105.00	237	187.07	756	242.57
91	101.70	246	176.13	768	232.12
91	102.90	258	183.40	860	246.70
97	110.00	276	186.26		

Source: Dudzinski and Mykytowycz, 1961.

As the second observation was selected using the table of random numbers, the sequence thus starts with $(X_1 = 29, Y_1 = 40.55)$ and terminates with $(X_9 = 768, Y_9 = 232.12)$.

Table 6.2 gives the LS estimates of the parameters of (6.2) for these two reduced data sets, together with their associated t-values. As expected, the t-values become smaller as the sample size is progressively reduced. Values for the curvature measures of nonlinearity and for bias increase as the sample size becomes progressively smaller; the parameter-effects nonlinearity is statistically significant with 9 data observations, but with 18

Table 6.2 Parameter Estimates and Associated t-Values for Model Function (6.2) in Combination with the Data of Table 6.1

Parameter	71 Observations		18 Observations		9 Observations	
	Estimate	t-value	Estimate	t-value	Estimate	t-value
α	5.63991	282.5	5.61235	209.0	5.61175	105.7
β	130.584	22.8	119.522	16.6	115.414	7.0
γ	37.6029	16.2	32.9892	11.4	31.0400	3.9

Table 6.3 Box's Percentage Bias and Intrinsic (IN) and Parameter-Effects (PE) Curvatures, for Model Function (6.2) in Combination with the Data of Table 6.1

Parameter	71 Observations	18 Observations	9 Observations
α	.005%	.009%	.037%
β	.120%	.220%	1.130%
γ	.157%	.298%	1.803%
IN	.036	.049	.095
PE	.163	.210	.545
Critical value $1/(2\sqrt{F})$	.302	.276	.229

observations the behavior is still sufficiently close to linear. The effect of reducing the number of observations from 18 to 9 is far more drastic than the effect of reducing the observations from 71 to 18, and illustrates the difficulty of trying to achieve a close approach to linear behavior in a model/data set combination for which the residual degrees of freedom is very small.

Simulation studies (Sec. 2.6) consisting of 1000 trials each were also carried out, based on the 18 and the 9 observations. The results for sample size 18 (Table 6.4) show that the distribution of the set of 1000 LS estimates of each parameter does not deviate significantly

Table 6.4 Results of Simulation Studies for Model Function (6.2) for
Reduced Sample Sizes

	Sample size 18	Sample Size 9
α		
% Bias	$.005^{ns}$	$.062^{*}$
% Excess variance	2.119^{ns}	$-.275^{ns}$
Skewness	$.128^{ns}$	$.244^{**}$
Excess kurtosis	$.011^{ns}$	$.104^{ns}$
β		
% Bias	$.207^{ns}$	1.436^{**}
% Excess variance	$.379^{ns}$	5.345^{ns}
Skewness	$.083^{ns}$	$.530^{**}$
Excess kurtosis	$-.074^{ns}$	$.590^{**}$
γ		
% Bias	$.389^{ns}$	2.041^{*}
% Excess variance	$.796^{ns}$	5.925^{ns}
Skewness	$.052^{ns}$	$.487^{**}$
Excess kurtosis	$-.250^{ns}$	$.465^{**}$

Note: $^{ns}P > .05$; $^{*}P < .05$; $^{**}P < .01$.

from a normal distribution. In all cases the biases are small and the
variances of the estimates are close to the minimum variance bound.
For sample size 9, however, the distributions of the estimates of each
parameter do deviate from normality. The negative excess variance for
parameter α is a chance effect for this particular simulation study,
as obviously an excess variance should not be negative. The skewness
coefficient is highly significant ($P < .01$) for each parameter,
indicating significant lack of symmetry and hence lack of linearity.

The percentage biases obtained from simulation (Table 6.4) show
excellent agreement in order of magnitude with the calculated biases
using Box's formula (2.3) (Table 6.3).

6.3 Model Describing Catalytic Chemical Reaction

A common problem in studies of chemical reaction is to describe the
relationship between the rate of reaction and the concentration of the
reactants. The model function under consideration is

$$E(Y) = \frac{\theta_1 \theta_3 X_1}{1 + \theta_1 X_1 + \theta_2 X_2} \tag{6.3}$$

There are two regressor variables X_1 and X_2, in contrast to all other illustrative problems considered previously. The data set for use in conjunction with (6.3) is taken from a paper of Meyer and Roth (1972, example 1) and involves the catalytic dehydration of n-hexyl alcohol, a reaction that had previously been considered by Box and Hunter (1965). The data, given in Table 6.5, are very sparse, there being only 5 observations.

The LS estimates of the parameters of (6.3) for the data of Table 6.5 are given in Table 6.6 together with their associated t-values. The relatively high t-value for parameter θ_2 suggests that this parameter is well determined, but parameters θ_1 and θ_3 appear to be less well determined. The intrinsic curvature for this model/data set

Table 6.5 Rate of Reaction Y versus Partial Pressure of Alcohol X_1 and Olefin X_2

X_1	X_2	Y
1.0	1.0	.126
2.0	1.0	.219
1.0	2.0	.076
2.0	2.0	.126
.1	.0	.186

Source: Data from Meyer and Roth, 1972.

Table 6.6 Parameter Estimates and Associated t-Values for Model Function (6.3) in Combination with the Data of Table 6.5

Parameter	LS estimate	t-value
θ_1	3.13	3.87
θ_2	15.16	24.00
θ_3	.78	5.13

combination is .038 and the parameter-effects curvature is 12.836; the critical value corresponding to the significance level α = .05 is .114. Hence, although this model function/data set combination has an acceptable intrinsic nonlinearity, its parameter-effects nonlinearity is highly significant. Therefore, it seems desirable to search for a reparameterization. A simulation study was carried out based on 1000 sets of data generated under the assumption that the LS estimates of θ_1, θ_2, and θ_3 given in Table 6.6 are the "true" values of the parameters, and that the observed residual variance $\hat{\sigma}_2$ = .00002179 is the "true" value of σ^2. The results show that the nonlinear behavior is due to parameters θ_1 and θ_3, especially the latter, the sampling distribution of whose estimates exhibits a long right-hand tail (Table 6.7). The biases and asymptotic covariance matrix for all parameters are presented in Table 6.8. There is relatively good agreement between the magnitudes of the simulated and calculated bias values.

A fairly obvious reparameterization of (6.3) is to replace the product of the two parameters $\theta_1\theta_3$ by a single new parameter while leaving the parameters in the denominator unchanged. The new parameters are related to the old parameters as follows:

$$\phi_1 = g_1(\underset{\sim}{\theta}) = \theta_1\theta_3$$
$$\phi_2 = g_2(\underset{\sim}{\theta}) = \theta_1 \qquad\qquad (6.4)$$
$$\phi_3 = g_3(\underset{\sim}{\theta}) = \theta_2$$

To obtain a preliminary indication of the behavior of the new parameter ϕ_1, we make use of formula (2.5), which, for a function $g(\underset{\sim}{\theta})$ of two parameters, reduces to the following terms (see Exercise 2.2):

$$\text{Bias }(\hat{\phi}) = \text{Bias }(\hat{\theta}_1)\,\frac{\partial g(\underset{\sim}{\theta})}{\partial\theta_1} + \text{Bias }(\hat{\theta}_3)\,\frac{\partial g(\underset{\sim}{\theta})}{\partial\theta_3}$$
$$+ \frac{1}{2}\,\text{Var }(\hat{\theta}_1)\,\frac{\partial^2 g(\underset{\sim}{\theta})}{\partial\theta_1^2} + \frac{1}{2}\,\text{Var }(\hat{\theta}_3)\,\frac{\partial^2 g(\underset{\sim}{\theta})}{\partial\theta_3^2}$$
$$+ \text{Cov }(\hat{\theta}_1\hat{\theta}_3)\,\frac{\partial^2 g(\underset{\sim}{\theta})}{\partial\theta_1\,\partial\theta_3}$$

Table 6.7 Results of a Simulation Study of Model Function (6.3) Using the Parameter Estimates of Table 6.6

	Parameter		
	θ_1	θ_2	θ_3
% Bias	2.480^{**}	$.125^{ns}$	3.594^{**}
% Excess variance	4.257^{ns}	5.163^{ns}	38.671^{**}
Skewness	$.383^{**}$	$.098^{ns}$	1.452^{**}
Excess kurtosis	$.405^{**}$	$.255^{ns}$	3.633^{**}

Note: $^{ns}P > .05$; $^{**}P < .01$.

Table 6.8 Calculated Box Bias, Percentage Bias, and Asymptotic Covariance Matrix for Model Function (6.3) in Combination with Data in Table 6.5

Parameter	Bias	Percentage Bias	Covariance Matrix		
			θ_1	θ_2	θ_3
θ_1	.04610	1.473	.65345	.14663	-.12193
θ_2	.02137	.141	.14663	.39900	-.01747
θ_3	.03038	3.895	-.12193	-.01747	.02310

The partial derivatives, which are to be evaluated at the LS estimates $\hat{\underset{\sim}{\theta}}$, are

$$\frac{\partial g(\underset{\sim}{\theta})}{\partial \theta_1} = \theta_3$$

$$\frac{\partial g(\underset{\sim}{\theta})}{\partial \theta_3} = \theta_1$$

$$\frac{\partial^2 g(\underset{\sim}{\theta})}{\partial \theta_1^2} = \frac{\partial^2 g(\underset{\sim}{\theta})}{\partial \theta_3^2} = 0$$

$$\frac{\partial^2 g(\theta)}{\partial \theta_1 \, \partial \theta_3} = 1$$

Substituting values for the LS estimates of the parameters from Table 6.6 and biases, variances, and covariances from Table 6.8, the following is obtained:

$$\text{Bias } (\hat{\phi}) = .04610(.78) + .03038(3.13) + \frac{1}{2} (.65435)(0)$$

$$+ \frac{1}{2} (.02310)(0) + (-.12193) (1)$$

$$= .00911$$

As $\hat{\phi}_1 = \hat{\theta}_1 \hat{\theta}_3 = (3.13)(.78) = 2.44$, the percentage bias in $\hat{\phi}_1$ is

$$\% \text{ Bias } (\hat{\phi}_1) = \frac{100(.00911)}{2.44} = .373\%$$

or about a tenth of the percentage bias in $\hat{\theta}_3$.

Thus the reparameterization given by (6.4) may result in an improvement in the linear behavior of the model/data set combination. To examine this, the curvature measures of nonlinearity were calculated for the new model function resulting from the reparameterization, namely for

$$E(Y) = \frac{\phi_1 X_1}{1 + \phi_2 X_1 + \phi_3 X_2} \tag{6.5}$$

The intrinsic curvature is .038 as before, as reparameterization does not alter the position of the solution locus, but the parameter-effects curvature drops markedly to a value of .26, which is a considerable improvement over the original parameterization. However, as the critical value $1/(2\sqrt{F}) = .114$, the nonlinearity is still significant statistically. We thus search for some other reparameterization.

Consider the following model function, which is a fairly obvious reparameterization of (6.3),

$$E(Y) = \frac{X_1}{\phi_1 + \phi_2 X_1 + \phi_3 X_2} \tag{6.6}$$

Comparing (6.6) with (6.3), the two sets of parameters are related as follows:

$$\phi_1 = g_1(\underset{\sim}{\theta}) = \frac{1}{\theta_1 \theta_3}$$

$$\phi_2 = g_2(\underset{\sim}{\theta}) = \frac{1}{\theta_3}$$

$$\phi_3 = g_3(\underset{\sim}{\theta}) = \frac{\theta_2}{\theta_1 \theta_3}$$

The percentage biases for the new parameters are as follows (see Exercise 6.2):

% Bias $(\hat{\phi}_1)$ = .113%

% Bias $(\hat{\phi}_2)$ = -.099%

% Bias $(\hat{\phi}_3)$ = .093%

These percentage biases are all of the order of only .1%, suggesting that model function (6.6) may be behaving more linearly with respect to estimation than either model functions (6.3) or (6.5). The parameter-effects nonlinearity is .089, which is less than the critical value $1/(2\sqrt{F})$ = .114; this further supports the notion of a

Table 6.9 Results of a Simulation Study for Model Function (6.6) Based on Parameter Estimates and Residual Variance Derived from the Data Set of Table 6.5

	Parameter		
	ϕ_1	ϕ_2	ϕ_3
% Bias	$-.104^{ns}$	$.641^{ns}$	$-.151^{ns}$
% Excess variance	1.981^{ns}	$-.512^{ns}$	$-.456^{ns}$
Skewness	$.033^{ns}$	$-.045^{ns}$	$.171^{*}$
Excess kurtosis	$-.016^{ns}$	$-.067^{ns}$	$-.092^{ns}$

Note: $^{ns}p > .05$; $^{*}p < .05$.

close-to-linear behavior of the LS estimates of the parameters of (6.6). To test the validity of this conclusion, a simulation study consisting of 1000 trials was carried out (Table 6.9). Except for the significant ($P < .05$) skewness coefficient for the estimates of parameter ϕ_3, which may be a chance effect, the behavior of the LS estimators of all parameters in model function (6.6) is seen to be close to linear with respect to parameter estimation.

6.4 Model and Data Set from a Paper of Meyer and Roth (1972)

The model and data set for this section are taken from a paper of Meyer and Roth (1972, example 7). They do not state the process, mechanism, experiment, or system concerned. There are 10 observations, and these are given in Table 6.10.

The model function under consideration to describe these data is

$$E(Y) = \theta_1 + \theta_2 \exp(\theta_3 X) \tag{6.7}$$

This is the same model function as model function (5.2) of the asymptotic regression model considered in Chapter 5, except for the signs of θ_2 and θ_3.

Table 6.10 Data from Unspecified Experiment

X	Y
1	16.7
5	16.8
10	16.9
15	17.1
20	17.2
25	17.4
30	17.6
35	17.9
40	18.1
50	18.7

Source: Data from Meyer and Roth, 1972.

The LS estimates of the parameters of (6.7) for the data of Table 6.10 are given by Meyer and Roth (1972) to be $\hat{\theta}_1 = 15.67$, $\hat{\theta}_2 = .999$, and $\hat{\theta}_3 = .022$ with the residual variance $\hat{\sigma}^2$ of .00131.

Using the estimates above, $\hat{\theta} = (\hat{\theta}_1, \hat{\theta}_2, \hat{\theta}_3)$ and $\hat{\sigma}^2$, the curvature measures of nonlinearity (Sec. 2.4) and bias calculation (Sec. 2.5) are given in Table 6.11. The intrinsic nonlinearity is acceptable, indicating that the shape of the solution locus is sufficiently close to that of a hyperplane, but the parameter-effects nonlinearity is unacceptable, so that at least one parameter in model function (6.7) exhibits nonlinear behavior in estimation. Results of a simulation study of 1000 trials (Sec. 2.6) are presented in Table 6.12. The estimator $\hat{\theta}_3$ exhibits close-to-linear behavior, but the distributions of $\hat{\theta}_1$ and $\hat{\theta}_2$ are highly non-normal. The distribution of $\hat{\theta}_1$ is negatively skewed with a long left-hand tail; conversely, that of $\hat{\theta}_2$ is positively skewed with a long right-hand tail. This suggests that the parameter θ_2 might be replaced by its exponential to give the model function

$$E(Y) = \theta_1 + \exp(\theta_4 + \theta_3 X) \tag{6.8}$$

whereas θ_1 might be replaced by its logarithm to give the model function

Table 6.11 Box's Percentage Bias and Intrinsic (IN) and Parameter-Effects (PE) Curvatures, for Model Function (6.7) in Combination with the Data of Table 6.10

Parameter	Model function (6.7)
θ_1	-.165%
θ_2	2.566%
θ_3	.027%
IN	.039
PE	7.578
Critical value $1/(2\sqrt{F})$	.240

$$E(Y) = \log \theta_5 + \exp (\theta_4 + \theta_3 X) \qquad\qquad (6.9)$$

The parameter-effects nonlinearity, although lower for model function (6.8) than for model function (6.7), and although lower again for model function (6.9) (Table 6.13), is nevertheless highly significant. The percentage biases do not in this case provide the modeler with any useful information about the behavior of the parameters in estimation.

Table 6.12 Results of a Simulation Study of Model Function (6.7) Using the Parameter Estimates and Residual Variance Obtained for the Data of Table 6.10

	Parameter		
	θ_1	θ_2	θ_3
% Bias	$-.222^{**}$	3.407^{**}	$-.330^{ns}$
% Excess variance	24.826^{**}	26.347^{**}	6.265^{ns}
Skewness	$-.863^{**}$	$.951^{**}$	$.030^{ns}$
Excess kurtosis	1.701^{**}	2.000^{**}	$.163^{ns}$

Note: $^{ns}P > .05;$ $^{**}P < .01.$

Table 6.13 Box's Percentage Bias and Intrinsic (IN) and Parameter-Effects (PE) Curvatures for Model Functions (6.8) and (6.9) for the Data Set of Table 6.10

Parameter	Model function (6.8)	Model function (6.9)
θ_1	$-.165\%$	–
θ_2	–	–
θ_3	$.027\%$	$.027\%$
θ_4	-608.680%	-608.680%
θ_5	–	$-.213\%$
IN	$.039$	$.039$
PE	4.650	2.961
Critical value $1/(2\sqrt{F})$	$.240$	$.240$

Table 6.14 Results of Simulation Study for Two of the Parameters in Model Function (6.9) Based on Data in Table 6.10

	Parameter	
	θ_4	θ_5
% Bias	-1252^{*}	-1.579^{*}
% Excess variance	10.408^{*}	$-.962^{ns}$
Skewness	$.196^{*}$	$.100^{ns}$
Excess kurtosis	$.315^{*}$	$-.059^{ns}$

Note: $^{ns}p > .05$; $^{*}p < .05$.

The extraordinary high percentage bias of -608.7% for $\hat{\theta}_4$ is due to the fact that $\hat{\theta}_4 = \log(\hat{\theta}_2) = \log(.999) = -.0010$; the high percentage bias occurs simply because the bias has been divided by a number close to zero. Results of a simulation study of 1000 trials for model function (6.9) are presented in Table 6.14. The LS estimators of θ_4 and θ_5 in model function (6.9) behave much more close to linear than the LS estimators of θ_1 and θ_2 in (6.7), but there is still evidence of some non-normal behavior, especially for $\hat{\theta}_4$.

6.5 Model Relating the Resistance of a Thermistor to Temperature

The model and data for this problem come from a paper of Meyer and Roth (1972, example 8); the models and data used in Sections 6.3 and 6.4 were also obtained from this paper. The data are reproduced in Table 6.15, and purport to represent the resistance of a thermistor Y to temperature X. However, as the resistance of a thermistor increases with temperature, Y probably represents conductance, which is inversely related to resistance.

Meyer and Roth assumed an additive stochastic term, so that the model function proposed by them to describe these data is

$$E(Y) = \theta_1 \exp\left(\frac{\theta_2}{X + \theta_3} \right) \tag{6.10}$$

It is left to readers as an exercise to show that the LS estimates of this model/data set combination are as follows: $\hat{\theta}_1$ = .00560964, $\hat{\theta}_2$ = 6181.35, $\hat{\theta}_3$ = 345.224, with $\hat{\sigma}^2$ = 6.769 (see Exercise 6.4).

Rather than examine this additive error case, which was also studied by Bates and Watts (1980), we will examine the multiplicative error case, which appears to be more appropriate as the residual variance tends to increase with increasing Y; the model function is

$$E(\log Y) = -\theta_1 + \frac{\theta_2}{X + \theta_3} \qquad (6.11)$$

A minus sign has been placed in front of θ_1 so that the LS estimate of this parameter will be positive for the data of Table 6.15. Except for the signs of θ_1 and θ_2, this model function is identical with model function (6.2) which related the age of wild rabbits to the weight of their eye lenses (Sec. 6.2).

The LS estimates of the parameter of model function (6.11) in combination with the data set of Table 6.15 are given in Table 6.16 together with their associated t-values. The very high t-values indicate that all parameters appear to be very well determined. Results for the curvature measures of nonlinearity (Sec. 2.4) and bias calculation (Sec. 2.5) are presented in Table 6.17.

Table 6.15 Data Set Representing the Conductance of a Thermistor (Y) as a Function of Temperature (X)

X	Y	X	Y
50	34,780	90	8,261
55	28,610	95	7,030
60	23,650	100	6,005
65	19,630	105	5,147
70	16,370	110	4,427
75	13,720	115	3,820
80	11,540	120	3,307
85	9,744	125	2,872

Source: Data from Meyer and Roth, 1972.

The intrinsic curvature is almost zero, indicating that the shape of the solution locus is very close to that of a hyperplane. The percentage biases in all three parameter estimates are also extremely low, suggesting close-to-linear behavior of the LS estimators for all parameters. This is not borne out, however, by the parameter-effects curvature value of 1.61 in Table 6.17, which is significantly high compared with the critical value $1/(2\sqrt{F})$ = .271 for assessing nonlinearity at α = .05. In an attempt to shed more light on the question, a simulation study consisting of 1000 trials was carried out on model (6.11) using the parameter estimates and residual variance derived from the data set of Table 6.15. The results of this simulation

Table 6.16 LS Estimates and Associated t-Values for Model Function (6.11) in Combination with the Data of Table 6.15

Parameter	LS estimate	t-value
θ_1	5.14483	303.9
θ_2	$.614878(10^4)$	422.6
θ_3	$.344107(10^3)$	478.3

Table 6.17 Box's Percentage Bias and Intrinsic (IN) and Parameter-Effects (PE) Curvatures, for Model Function (6.11) in Combination with the Data of Table 6.15

Parameter	Model function (6.11)
θ_1	.0004%
θ_2	.0004%
θ_3	.0002%
IN	.0002
PE	1.6056
Critical value $1/(2\sqrt{F})$	.271

study showed close-to-linear behavior of the LS estimators of each of the three parameters considered separately.

A further simulation study was carried out on the model/data set combination as described in Sec. 2.9 to estimate the asymmetry measure of bias. For all three parameters, the bias in the estimators obtained using (2.16) was extremely low, and the correlation coefficients using (2.18) of -.999995, -.999988, and -.999995, respectively, indicate close-to-linear behavior of each estimator. There are two possible explanations for what appears to be contradictory results. Either the parameter-effects curvature measure may sometimes indicate significant nonlinearity when none is present, or the curvature measure may be the more correct indicator as it operates in a p-dimensional space whereas simulation studies and the bias calculation of Box relate only to the marginal distribution of the estimators. Because of the lack of an applicable methodology for examining multidimensional space, the apparent contradiction cannot be easily resolved. The results for the multiplicative error case considered above, where the parameter-effects curvature is surprisingly high without any corresponding indication of nonlinear behavior from the bias calculation or simulation studies, is paralleled by the additive error case using model function (6.10). The demonstration of this is left to the reader (see Exercise 6.5).

6.6 Bent-Hyperbola Regression Models

We now consider models for data which appear to follow two distinctly different linear relationships, but allow for a smooth transition from one linear régime to the other. We first look at a model proposed by Griffiths and Miller (1973), referred to as a "bent-hyperbola" regression model, having the model function

$$E(Y) = \theta_1 + \theta_2(X - \theta_4) + \theta_3[(X - \theta_4)^2 + \theta_5]^{1/2} \qquad (6.12)$$

Three sets of data in combination with this model function will be examined in detail; these data are given in Table 6.18. The first data set, obtained from the unpublished Ph.D. thesis of R. A. Cook, Queen's

Table 6.18 Data for Bent-Hyperbola Models

Data set 1[a]		Data set 2[b]		Data set 3[b]	
X	Y	X	Y	X	Y
-1.39	1.12	1	290.426	1	113.978
-1.39	1.12	2	295.632	2	115.288
-1.08	.99	3	299.183	3	116.709
-1.08	1.03	4	302.900	4	117.624
-.94	.92	5	307.454	5	118.776
-.80	.90	6	311.313	6	120.692
-.63	.81	7	316.910	7	121.207
-.63	.83	8	321.523	8	122.764
-.25	.65	9	325.605	9	124.405
-.25	.67	10	329.894	10	125.280
-.12	.60	11	333.469	11	126.396
-.12	.59	12	336.373	12	127.867
.01	.51	13	339.984	13	129.172
.11	.44	14	342.868	14	130.354
.11	.43	15	344.641	15	131.483
.11	.43	16	343.848	16	133.038
.25	.33	17	345.530	17	133.497
.25	.30	18	346.104	18	135.159
.34	.24	19	345.197	19	135.099
.34	.25	20	346.024	20	135.787
.44	.13	21	345.731	21	136.853
.59	-.01	22	346.648	22	136.057
.70	-.13	23	345.517	23	137.150
.70	-.14	24	346.544	24	137.145
.85	-.30	25	346.899	25	137.162
.85	-.33			26	137.724
.99	-.46			27	137.128
.99	-.43				
1.19	-.65				

[a] X = log (flow rate in g/cm sec); Y = log (band height in cm).

[b] X = time intervals; Y = radioactivity counts/1000.

Source: Data set 1 from Bacon and Watts, 1971; data sets 2 and 3 courtesy of A. Lang, CSIRO Division of Plant Industry, Canberra, Australia.

University, Kingston, Ontario, and used by Bacon and Watts (1971), involves the behavior of the stagnant surface layer height as a function

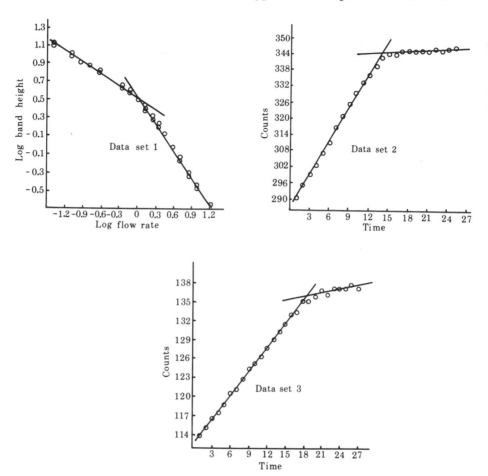

Fig 6.2 Graph of data (Table 6.18) for bent-hyperbola models.

of the controlled flow of water down an inclined channel. The second
and third data sets were obtained from A. Lang, CSIRO Division of Plant
Industry, Canberra, and involves sap flow in the long leaf stalk of
a water lily (*Nymphoides peltata*) which was monitored by applying a
radioactive tracer to the sap near the leaf, the source of flow, and
measuring its accumulation in the rootstock sink. Flow slowed rapidly
when the stalk was chilled. The three data sets are shown graphically
in Fig. 6.2. The LS estimates of the parameters of model function (6.12)

Table 6.19 LS Estimates and Associated t-Values for Model Function (6.12) in Combination with the Data of Table 6.18

Parameter	Data set 1		Data set 2		Data set 3	
	Estimate	t-value	Estimate	t-value	Estimate	t-value
θ_1	.58597	16.33	346.371	301.61	136.822	142.17
θ_2	-.73508	-56.86	2.23911	46.87	.695993	15.41
θ_3	-.35909	-14.38	-2.18994	-25.46	-.586726	-9.47
θ_4	.062696	2.02	13.5471	75.63	18.7515	36.47
θ_5	.096180	1.82	4.59719	2.10	6.72716	1.20

Table 6.20 Box's Percentage Bias and Intrinsic (IN) and Parameter-Effects (PE) Curvatures for Model Function (6.12) in Combination with the Data of Table 6.18

	Data set		
Parameter	1	2	3
θ_1	.594%	.029%	.124%
θ_2	.099%	-.027%	-.996%
θ_3	.721%	.296%	1.761%
θ_4	2.131%	.012%	.330%
θ_5	9.679%	6.595%	20.521%
IN	.308	.232	.434
PE	1.507	1.006	1.812
Critical value $1/(2\sqrt{F})$	.309	.304	.307

for the data in Table 6.18 are given in Table 6.19 together with their associated t-values. The low t-values associated with the estimates of θ_5 indicate that this parameter is not well determined in any of the three data sets. In addition, the estimates of θ_4 for data set 1 have a relatively low t-value.

Results for the curvature measures of nonlinearity of Bates and Watts (Sec. 2.4) and the bias calculation of Box (Sec. 2.5) are presented in Table 6.20. The values of the intrinsic curvature are relatively high, being significant for data set 3 and being just at the point of significance for data set 1, indicating that the solution locus is not well approximated by a hyperplane. The significant parameter-effects curvatures for all data sets indicate that at least one of the LS estimators is exhibiting nonlinear behavior. Results for the various derived measures from a simulation study of 1000 trials are presented in Table 6.21 for data set 2, following the procedure detailed in Sec. 2.6. These results show that the most nonlinear-behaving parameter is θ_5; a histogram of its LS estimates, which clearly demonstrates lack of normality, is shown in Fig. 6.3a. The second most nonlinear-behaving parameter is θ_1, but its behavior was not evident either from its t-value (Table 6.19) or from the percentage bias (Table 6.20). This is because the parameter θ_1 represents a "constant" term in the model function (6.12). The magnitude of the estimate $\hat{\theta}_1$ may be altered simply by a change in location of the response variable Y. By adding a constant to each Y value, the standard error and bias (i.e., absolute bias, *not* percentage bias) associated with the LS estimator remains unaltered, but the

Table 6.21 Results of a Simulation Study for Model Function (6.12) Using the Parameter Estimates and Residual Variance Estimate Obtained for Data Set 2 of Table 6.18

	\multicolumn{5}{c}{Parameter}
	θ_1	θ_2	θ_3	θ_4	θ_5
% Bias	$.035^{**}$	$-.069^{ns}$	$-.316^{*}$	$.075^{ns}$	6.312^{**}
% Excess variance	11.646^{*}	5.218^{ns}	4.015^{ns}	4.503^{ns}	20.169^{**}
Skewness	$.845^{**}$	$-.125^{ns}$	$-.665^{**}$	$.119^{ns}$	1.223^{**}
Excess kurtosis	1.556^{**}	$-.137^{ns}$	$.798^{**}$	$-.101^{ns}$	2.280^{**}

Note: $^{ns}P > .05$; $^{*}P < .05$; $^{**}P < .01$.

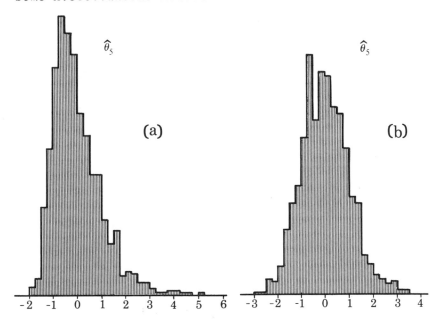

Fig 6.3 Histogram of results of simulation study for $\hat{\theta}_5$ in (a) model function (6.12); (b) model function (6.13).

t-value and percentage bias changes. In the extreme case, by a location change which makes $\sum_i Y_i = 0$, one would obtain an infinite t-value and an infinite percentage bias. Because of this, a simulation study appears to be unavoidable if one wishes to identify the nonlinear-behaving parameters in model functions such as (6.12) which have a "constant" term.

We now attempt to find a suitable reparameterization for θ_5 using data set 2. Since the simulation study showed the LS estimates of θ_5 to be skewed to the right (Fig. 6.3a), we can try replacing θ_5 by the exponential of a new parameter or by a new parameter raised to a positive power. A preliminary examination of the effect of these alternative reparameterizations may be made by using formulas (2.7) and (2.8). We will require the bias and asymptotic variance of $\hat{\theta}_5$ from model function (6.12) and these are given as follows, the subscript of θ_5 being dropped to simplify the notation:

$$\hat{\theta} = 4.59719$$

$$\text{Var } (\hat{\theta}) = 4.77029 \quad [\text{from the asymptotic covariance matrix}]$$

$$\text{Bias } (\hat{\theta}) = 0.303192 \quad [\% \text{ bias} = \frac{100(.303192)}{4.59719} = 6.60\%]$$

(i) Consider replacing θ by its exponential; that is, the new parameter will be related to the old θ by the following expression:

$$g(\theta) = \log \theta$$

Therefore,

$$g(\hat{\theta}) = \log \hat{\theta} = 1.52545$$

Further,

$$\frac{\partial g(\theta)}{\partial \theta} = \frac{1}{\hat{\theta}} = .217524$$

$$\frac{\partial^2 g(\theta)}{\partial \theta^2} = \frac{-1}{\hat{\theta}^2} = -.047317$$

From formula (2.7),

$$\text{Bias } [g(\hat{\theta})] = \text{Bias } (\hat{\theta}) \frac{\partial g(\theta)}{\partial \theta} + \frac{1}{2} \text{Var } (\hat{\theta}) \frac{\partial^2 g(\theta)}{\partial \theta^2}$$

$$= .303192(.217524) + \frac{1}{2} (4.77029)(-.047317)$$

$$= .06595 - .11286 = -.04691$$

$$\% \text{ Bias } [g(\hat{\theta})] = \frac{100(-.04691)}{1.52545} = -3.07\%$$

From formula (2.8),

$$\text{Var } [g(\hat{\theta})] = \left[\frac{\partial g(\theta)}{\partial \theta} \right]^2 \text{Var } (\hat{\theta})$$

$$= (.217524)^2 (4.77029) = .2257$$

(ii) Consider replacing θ by its square; that is, the new parameter will be related to the old θ by the following expression:

$$g(\theta) = \theta^{1/2}$$

Therefore,

$$g(\hat{\theta}) = \hat{\theta}^{1/2} = 2.14411$$

Further,

$$\frac{\partial g(\theta)}{\partial \theta} = \frac{\hat{\theta}^{-1/2}}{2} = .233197$$

$$\frac{\partial^2 g(\theta)}{\partial \theta^2} = \frac{-\hat{\theta}^{-3/2}}{4} = -.025363$$

From formula (2.7),

$$\text{Bias } [g(\hat{\theta})] = .303192(.233197) + \frac{1}{2}(4.77029)(-.025363)$$

$$= .070703 - .060494 = .010209$$

$$\% \text{ Bias } [g(\hat{\theta})] = \frac{100(.010209)}{2.14411} = .48\%$$

From formula (2.8),

$$\text{Var } [g(\hat{\theta})] = (.233197)^2 (4.77029) = .2594$$

The results of the calculations above indicate that replacing the parameter θ_5 by its exponential tends to overcompensate; whereas originally the bias was positive, the new bias is negative ($-.04691$). In a simulation study of 1000 trials, the bias is likely to be adjudged to be significantly negative, as

$$\frac{\text{Bias } [g(\hat{\theta})]}{\{\text{Var } [g(\hat{\theta})]/1000\}^{1/2}} = \frac{-.04691}{(.2257/1000)^{1/2}} = -3.12$$

which is highly significant (P < .01) when referred to a table of the standard normal distribution.

On the other hand, replacing the original parameter θ_5 by its square results in only a small positive bias (.010209) in the new parameter, which, in a simulation study of 1000 trials, is likely not to be significant, as

$$\frac{\text{Bias } [g(\hat{\theta})]}{\{\text{Var } [g(\hat{\theta})]/1000\}^{1/2}} = \frac{.010209}{(.2594/1000)^{1/2}} = .63$$

which is not significant when referred to a table of the standard normal distribution.

Hence the model function

$$E(Y) = \theta_1 + \theta_2(X - \theta_4) + \theta_3[(X - \theta_4)^2 + \theta_5^2]^{1/2} \tag{6.13}$$

may provide a more suitable parameterization than model function (6.12) insofar as parameter θ_5 is concerned. This same model function, except for the term θ_5^2 being written as $\theta_5^2/4$, was considered by Watts and Bacon (1974). For this model function, the curvature measures of nonlinearity and the bias calculation of Box for all data sets of Table 6.18 are presented in Table 6.22. Only the percentage biases in parameter θ_5 and the parameter-effects curvatures are listed in Table 6.22, as the LS estimates of the other parameters, the statistical behavior thereof, and the intrinsic curvatures are the same as for model function (6.12). In general, reparameterization affects only the parameter-effects curvature and the LS estimates and behavior of the reparameterized parameters.

The percentage biases in the LS estimates of θ_5 (in the form θ_5^2) shown in Table 6.22 are much lower than those for θ_5 in Table 6.20, and simulation studies of 1000 trials confirm that the parameterization of θ_5 in model function (6.13) is more close to linear in behavior than the parameterization of θ_5 in model function (6.12). The results of one of these simulation studies are shown in Table 6.23 for data set 2; although the skewness coefficient is significant, the departure from

Table 6.22 Box's Percentage Bias and Parameter-Effects (PE) Curvatures for Model Function (6.13) in Combination with the Data of Table 6.18

	Data set		
Parameter	1	2	3
θ_5	1.069%	.476%	1.609%
PE	3.716	1.964	4.479
Critical value $1/(2\sqrt{F})$	.309	.304	.307

normality is very slight, as can be seen from the histogram of $\hat{\theta}_5$ in Fig. 6.3b.

The parameter-effects curvatures are higher for model function (6.13) (Table 6.22) for all data sets than those for model function (6.12) (Table 6.20). This is surprising at first glance, as reparameterization of θ_5 results in improved linear behavior with respect to that parameter. However, it must be remembered that other parameters are behaving nonlinearly in model functions (6.12) and (6.13), and the curvature measure of Bates and Watts is a measure of the parameter-effects nonlinearity in an overall multiparameter sense. Improving the behavior in estimation of one parameter in a model function does not necessarily guarantee an improvement in the joint behavior in estimation of that parameter with another parameter or parameters. Furthermore, the parameter-effects curvature measure examines the parallelism and spacing of parameter lines on a *tangent plane* to the solution locus, rather than on the solution locus itself. For model functions (6.12) and (6.13), the solution locus either departs significantly from a hyperplane, or is close to departing significantly, for all data sets. Hence caution must be exercised in searching for reparameterizations in this situation. It may be better for the user to look for an entirely different model which, in combination with the data, has a solution locus with a more acceptable intrinsic curvature.

Table 6.23 Results for Parameter θ_5 from a Simulation Study of 1000 Trials for Model Function (6.13) Using the Values of X from Data Set 2 of Table 6.18 and the Parameter Estimates and Residual Variance Derived Therefrom

% Bias	$-.746^{ns}$
% Excess variance	5.411^{ns}
Skewness	$.294^{**}$
Excess kurtosis	$.217^{ns}$

Note: $^{ns}P > .05;$ $^{**}P < .01.$

Hence we now examine an alternative bent-hyperbola model proposed by Bacon and Watts (1971) given by the expression

$$E(Y) = \theta_1 + \theta_2(X - \theta_4) + \theta_3(X - \theta_4) \tanh\left(\frac{X - \theta_4}{\theta_5}\right) \qquad (6.14)$$

The LS estimates of the parameters of (6.14) for the data in Table 6.18 are given in Table 6.24 together with their associated t-values.

Results for the curvature measures of nonlinearity of Bates and Watts and the bias calculation of Box are presented in Table 6.25. The intrinsic and parameter-effects curvatures are somewhat less than for model function (6.12) for data sets 1 and 2, but are higher for data set 3.

The parameter-effects curvature of .603 for data set 2 is not greatly in excess of the critical value of .304, especially in comparison with the large parameter-effects curvatures obtained for some models studied earlier in this book. Simulation studies of 1000 trials were carried out for all data sets. The results for data set 2 show that the behavior of the estimators of all parameters was close to linear except for $\hat{\theta}_1$, which exhibits slight negative skewness. The estimators for data set 1 tend to have greater percentage biases, skewness and kurtosis than those for data set 2, but the departures from normality of the distributions, although significant statistically, are slight. The simulation study for data set 3 was not successful, as 33 of the 1000 trials failed to converge; the greater

Table 6.24 LS Estimates and Associated t-Values for Model Function (6.14) in Combination with the Data of Table 6.18

Parameter	Data set 1		Data set 2		Data set 3	
	Estimate	t-value	Estimate	t-value	Estimate	t-value
θ_1	.4839	21.11	341.268	767.84	135.199	301.43
θ_2	−.7276	−60.44	2.258	54.45	.695	11.64
θ_3	−.2871	−30.00	−1.823	−59.37	−.499	−10.88
θ_4	.0473	1.60	13.482	85.11	18.739	28.54
θ_5	.6222	7.28	5.098	10.33	6.494	4.44

Table 6.25   Box's Percentage Bias and Intrinsic (IN) and Parameter-Effects (PE) Curvatures for Model Function (6.14) in Combination with the Data of Table 6.18

Parameter	Data set		
	1	2	3
θ_1	-.175%	.004%	.069%
θ_2	.107%	-.026%	-2.051%
θ_3	.587%	.234%	2.829%
θ_4	3.170%	.011%	.706%
θ_5	.330%	.112%	1.956%
IN	.251	.208	.478
PE	.946	.603	3.007
Critical value $1/(2\sqrt{F})$	.309	.304	.307

difficulty of obtaining convergence compared to the first two data sets is consistent with the fact that the curvature measures of nonlinearity are higher for this data set.

Exercises

6.1 For the data in Table 6.1 use a random selection procedure to produce reduced sets of observations of sample size 18 and 9, respectively. (Use a completely random selection procedure, rather than a systematic selection procedure with a random starting point, as was employed in Sec. 6.2). For the reduced data sets in combination with model function (6.2), (a) determine the LS estimates, (b) calculate the bias measure of Box and the curvature measures of nonlinearity of Bates and Watts, and (c) carry out simulation studies of 1000 trials each. Do the inferences from your results conform with those obtained from the results in Tables 6.2-6.4?

6.2 Use formula (2.5) to calculate the bias measure of Box for the LS estimates of the reparameterized parameters ϕ_1, ϕ_2, and ϕ_3 in model function (6.6), making use of the LS estimates of θ_1, θ_2, and θ_3 in Table 6.6, and their biases and covariance matrix in Table 6.8. Convert your answers to percentage bias and check your results against those reported in Sec. 6.3.

6.3 Fit model function (6.7) to the data of Table 6.10 and obtain the LS estimates of the parameters and the residual variance to greater accuracy than that reported in the text of Sec. 6.4.

6.4 Fit model function (6.10), which assumes additive error, to the data of Table 6.15, and confirm that the LS estimates of the parameters and the residual variance are as given in Sec. 6.5.

6.5 Using the LS estimates and residual variance obtained for the model and data of Exercise 6.4, calculate the curvature measures of nonlinearity of Bates and Watts and the bias measure of Box and carry out a simulation study of 1000 trials. Enterprising readers may also wish to carry out a simulation study as described in Sec. 2.9 to estimate the asymmetry measure of bias (2.16) and the associated correlation coefficients using (2.18). In the user's main calling routine (XAMPLE in the Appendix), values of Y_t^- may be obtained from generated values of Y_t^+ by making use of the relationship

$$Y_t^- = Y_t^+ - 2 \ (Y_t^+ - \hat{Y}_t)$$

where $\hat{Y}_t$ is the predicted value of the dependent variable corresponding to X_t and may be obtained by calling subroutine EVAL. Confirm that although the parameter-effects curvature indicates significant nonlinearity, the bias measures and the simulation studies indicate close-to-linear behavior.

6.6 Carry out a simulation study of 1000 trials for model function (6.14) in combination with data set 2 of Table 6.18, using the parameter estimates of Table 6.24 and the residual variance $\sigma^2 = .308351$. Carry out the tests detailed in Sec. 2.6 to examine how closely the set of estimates for each parameter approaches that expected of a linear model. Are the results in conformity with indications given by the curvature measures and the percentage biases in Table 6.25?

7

Comparing Parameter Estimates from More Than One Data Set

7.1 Introduction

Chapters 3-6 were devoted to a study of various nonlinear regression models with attention being directed toward the examination of the behavior of the least-squares estimators of the parameters of the models. It is seldom that the modeler is concerned with a single set of data only. Often one wishes to examine whether changing experimental conditions are reflected in changes in the parameters of the model. It is of interest to the modeler to test whether different experimental conditions lead to changes in one or more of the model parameters. A general test approach is to use a procedure based on what is referred to by Draper and Smith (1981) as the "extra sum of squares" principle and by Milliken and DeBruin (1978) as the "conditional error" principle. The procedure will be illustrated in the following two sections; Sec. 7.2 deals with linear models and Sec. 7.3 with nonlinear models.

7.2 Comparing Parameter Estimates in Linear Models

Although we will be concerned mainly with the comparison of parameter estimates in nonlinear models, an initial consideration of linear models will provide the necessary background to the approach. No new principles are introduced with nonlinear models, and the computer program in the Appendix of this book for obtaining the LS estimates of the parameters may be employed for both linear and nonlinear models.

A common use of the so-called "comparison of regressions" procedure occurs with the straight-line regression model where the user, having several sets of data, may wish to test whether a single, common line is adequate to fit all data sets. If the hypothesis of a common line is rejected, a more restricted hypothesis that the lines all have a common slope, but different intercepts, may be tested. The formulas for carrying out these tests are given in such books as Williams (1959, Chap. 8) and Li (1964, Chap. 19). The procedure to be used here relies on exactly the same principle, although the details may appear to be different because of variations in the sequence in which the models are fitted.

The procedure is perhaps best illustrated by an example, and for this purpose the data on yield versus density for onions, given in Appendix 3.A, will be employed. These data are ideally suited to illustrate the techniques examined in this chapter as they have been obtained for two varieties each grown at two localities. Readers will recall from Chapter 3 that the parameter α of the Holliday model (3.2) can be interpreted as a genetic parameter; therefore, it is reasonable to test a null hypothesis which postulates that α may be invariant within a variety even if grown at different localities. Similarly, β can be interpreted as an environmental parameter, and therefore it is reasonable to expect that β may be almost invariant for the same variety grown in similar environments.

Each of the four sets of data in Appendix 3.A consists of 42 observations, in the form of pairs of yields Y and corresponding densities X. Graphs of Y versus X such as Fig. 3.1 assist the user in examining whether there are any apparent atypical observations in the data, that is, observations which are far from the fitted line and raise suspicions that an experimental error has been made. From such graphs, two observations stand out as atypical; these are marked in Appendix 3.A. The point (X = 120.89, Y = 26.32) from Uraidla seems obviously atypical; the point (X = 180.39, Y = 28.96) from Virginia is less obviously atypical, but the removal of both these points from the data set makes the following illustrations more conclusive. These two observations are not shown in Fig. 3.1, as they would have fallen outside the boundaries of the axes. In all that follows, these two points have been omitted from consideration. Hence the data consist of 166 observations, 42 each at Mount Gambier (MG) and Purnong Landing (PL), and 41 each at Uraidla (U) and Virginia (V).

To illustrate the method for comparing parameters in linear models, we use the following linearized form of the Holliday model with an additive error assumption:

$$E\left(\frac{1}{Y}\right) = \alpha + \beta X + \gamma X^2 \qquad (7.1)$$

This was obtained by taking the reciprocal of the yield Y in model (3.2), producing a model which is linear in the parameters α, β, and γ. As discussed in Chapter 3, it is usual in yield-density studies to find that Var (log Y) is closer to being constant than Var (Y). Similarly, the variance of 1/Y [i.e., Var (1/Y)] is generally observed to be more nearly constant than Var (Y). Gillis and Ratkowsky (1978) found in simulation studies that the performance of model (7.1) was quite satisfactory even when the data were generated assuming a stochastic term which had constant variance on the logarithmic scale. This demonstrates the inherent robustness of linear estimators to departures from the underlying assumptions about the error variance. In any event, model (7.1) provides a suitable means of illustrating the methodology of studying parameter invariance with a linear model, and in addition, the parameter estimates obtained may then be used as initial estimates for the nonlinear model to be investigated in Sec. 7.3.

A formal system for analyzing the data of Appendix 3.A will now be proposed. Graphs of the data in Fig. 3.1 suggest that a model of straight-line regression of 1/Y on X may be adequate. Hence the first step is to decide whether the term γX^2 in (7.1) is necessary. Previous experience with these data in Chapter 3 indicated that this term was unnecessary (although with a different assumed structure for the stochastic term) and that the data obeyed the so-called "asymptotic" yield-density relationship. We thus wish to test whether $\gamma = 0$, or alternatively, whether the model

$$E\left(\frac{1}{Y}\right) = \alpha + \beta X \qquad (7.2)$$

is adequate. To perform the test, the following two-step prescription may be used with model (7.1):

(A) Fit α, β, and γ to each individual set of data. As there are three parameters for each of four sets of data, a total of 12 parameters will be estimated, with the pooled residual degrees of freedom being $166 - 12 = 154$. Assuming that the residual variances are homogeneous, the residual sum of squares from each of the four sets of data are added together to produce a pooled residual sum of squares. As an alternative to fitting the four data sets separately, all 12 parameters may be estimated in a single estimation by minimizing

$$S_A = \sum_{i=1}^{4} \sum_{t=1}^{n_i} \left(\frac{1}{Y_t} - \alpha_{iA} - \beta_{iA}X_t - \gamma_{iA}X_t^2\right)^2 \qquad (7.3)$$

where n_i is the sample size of the ith data set (in this example 42, 41, 42, and 41, respectively, for MG, U, PL, and V). The parameter estimates $\hat{\alpha}_{iA}$, $\hat{\beta}_{iA}$, $\hat{\gamma}_{iA}$, $i = 1, 2, 3, 4$, will be identical to those obtained by fitting each individual set of data separately, and the residual sum of squares RSS_A will be identical to the sum of squares obtained by pooling the individual sum of squares.

(B) Fit α and β to each individual set of data. Ignoring γ is equivalent to the assumption $\gamma = 0$. A total of 8 parameters is estimated, with the pooled residual degrees of freedom being $166 - 8 = 158$. The residual sum of squares RSS_B may be obtained, as in (A), by pooling the residual sum of squares from the individual data sets, or by minimizing

$$S_B = \sum_{i=1}^{4} \sum_{t=1}^{n_i} \left(\frac{1}{Y_t} - \alpha_{iB} - \beta_{iB}X_t\right)^2 \qquad (7.4)$$

The parameter estimates $\hat{\alpha}_{iB}$, $\hat{\beta}_{iB}$, $i = 1, 2, 3, 4$, will be identical whichever approach is employed.

The results of using steps (A) and (B) are summarized in Table 7.1. The residual sum of squares RSS_B, from fitting step (B), is larger than RSS_A, from fitting step (A), since model (7.2) is a more restricted model than (7.1), the latter containing all the terms of the former model plus an additional term. The difference $RSS_B - RSS_A = 4.89(10^{-6})$, when divided by the difference in the degrees of freedom $158 - 154 = 4$, gives a mean square of $1.2225(10^{-6})$, which may be tested against the residual mean square derived from the more general model (7.1), namely $RSS_A/154 = 1.2428(10^{-6})$. This ratio,

$$F = \frac{1.2225(10^{-6})}{1.2428(10^{-6})} = .984$$

has an F-distribution with 4 and 154 degrees of freedom; hence γ is adjudged to be not significantly different from zero. This result confirms the conclusion obtained from Chapter 3 that the growth of onions obeys the asymptotic yield-density model.

Having now established that model (7.2) is adequate to describe the yield-density response for each of the four data sets of Appendix

Table 7.1 Comparison of Fits for Four Onion Data Sets of Appendix 3.A with Parameter γ Included and Excluded, for Linear Model (7.1) Assuming Var (1/Y) Constant

Description of fit or test	Number of parameters estimated, p	Degrees of freedom, df	Residual sum of squares, RSS	Residual mean square, RMS
(B) Individual α, $\beta(\gamma = 0)$	8	158	.00019628	
(A) Individual α, β, γ	12	154	.00019139	$1.2428(10^{-6})$

		df	Change in RSS	Mean square
(B) $-$ (A): test of $\gamma = 0$		4	.00000489	$1.2225(10^{-6})$

3.A, we now turn attention to such questions as to whether α and/or β may be invariant. As part of the general prescription to perform these tests, the following three steps are required:

(C) Fit a common α and β to the four data sets. This involves simply minimizing the sum of squares

$$S_C = \sum_{i=1}^{4} \sum_{t=1}^{n_i} \left(\frac{1}{Y_t} - \alpha_C - \beta_C X_t \right)^2 \tag{7.5}$$

which is equivalent to fitting model (7.2) to the pooled 166 observations. Two parameters are estimated, with the pooled residual degrees of freedom being $166 - 2 = 164$.

(D) Fit a common α to the four data sets, allowing the β's to be different. This involves minimizing the sum of squares

$$S_D = \sum_{i=1}^{4} \sum_{t=1}^{n_i} \left(\frac{1}{Y_t} - \alpha_D - \beta_{iD} X_t \right)^2 \tag{7.6}$$

Five parameters are estimated, α_D, β_{1D}, β_{2D}, β_{3D}, and β_{4D}, with the pooled residual degrees of freedom being $166 - 5 = 161$.

(E) Fit a common β to the four data sets, allowing the α's to be different. This involves minimizing the sum of squares

$$S_E = \sum_{i=1}^{4} \sum_{t=1}^{n_i} \left(\frac{1}{Y_t} - \alpha_{iE} - \beta_E X_t \right)^2 \tag{7.7}$$

Five parameters are estimated, α_{1E}, α_{2E}, α_{3E}, α_{4E}, and β_E, with the pooled residual degrees of freedom being $166 - 5 = 161$.

The results of following the prescription above are summarized in Table 7.2. The entries corresponding to steps (A) and (B), which were also given in Table 7.1, are repeated in Table 7.2 to show the full sequence of tests. The difference (B) − (A) provided a test of whether $\gamma = 0$, and this hypothesis was not rejected. The next step is to test the acceptability of the most restricted model, which is that corresponding to step (C) and which postulates the invariance of both α and β over all four data sets. It is not absolutely clear whether this test should be made relative to step (A), which is based on the three-parameter Holliday model (7.1), or to step (B), based on the two-parameter asymptotic model (7.2). There are no clear-cut statistical guidelines for preferring either, as both the residual mean square (RMS) based on RSS_A and the RMS based on RSS_B are unbiased if (7.1) is the correct model and if the null hypothesis that $\gamma = 0$ in model (7.1) is true. For these data, the two RMS values are almost identical and the inferences will be the same. As it does not matter which RMS we use, we will calculate all variance ratios using the RMS based on RSS_A.

As the variance ratio for the test of invariance of α and β is 64.2 and therefore highly significant (F-distribution with 6 and 154 degrees of freedom, P < .01), the hypothesis of a common value of α and β must be rejected. Similarly, a test of the invariance of α (allowing β to vary from data set to data set), based on the difference (D) − (B), leads to a variance ratio of 8.11 (P < .01), and a test of

Table 7.2 Comparison of Fits for Onion Data of Appendix 3.A to Determine Whether Parameters α and/or β Are Invariant

Description of fit or test	Number of parameters estimated, p	Degrees of freedom df	Residual sum of squares, RSS	Residual mean square, RMS
(E) Common β	5	161	.00026295	
(D) Common α	5	161	.00022653	
(C) Common α and β	2	164	.00067536	
(B) Individual α and β	8	158	.00019628	$1.2423(10^{-6})$
(A) Individual α, β, and γ	12	154	.00019139	$1.2428(10^{-6})$

	df	Change in RSS	Mean square	Variance ratio, F
(B) − (A): test of $\gamma = 0$	4	.00000489	$1.2225(10^{-6})$	$.984^{ns}$
(C) − (B): test of invariance of α and β	6	.00047908	$79.847(10^{-6})$	64.2^{**}
(D) − (B): test of invariance of α	3	.00003025	$10.083(10^{-6})$	8.11^{**}
(E) − (B): test of invariance of β	3	.00006667	$22.223(10^{-6})$	17.88^{**}

Note: ^{ns}P > .05; **P < .01.

the invariance of β (allowing α to vary from data set to data set), based on the difference (E) − (B), leads to a variance ratio of 17.88 (P < .01), and therefore the hypothesis of either a common α or a common β must be rejected. The next step is to split the four data sets into two groups of two data sets, the first group being for the variety Brown Imperial Spanish grown at MG and U, and the second group being for the variety White Imperial Spanish grown at PL and V. It will be left to the reader as an exercise (see Exercise 7.4) to carry out the calculations for each of these two groups based on the linear model (7.2). As stated earlier, it is usual in yield-density work to find that log Y has a constant variance (or approximately so), and this assumption leads to a nonlinear model. In using the linear model (7.2) in this section, we have assumed that 1/Y has constant variance, which reasonably approximates the other assumption. The advantage of the linear model is not only its simplicity but also its guarantee of success in enabling the LS estimates of the parameters to be obtained. Using the Gauss–Newton algorithm, such as appears in the Appendix to this book, a linear model will converge to the LS estimates of the parameters in a single iteration from any set of initial parameter estimates (see Appendix 2.A).

The values of the parameter estimates for α and β from step (B), together with their standard errors obtained as the square root of the diagonal elements of the covariance matrix, are given in Table 7.3, the subscripts indicating the localities. Examination of these estimates and their standard errors suggests that the values of α may be invariant within each variety. In addition, the values of β for

Table 7.3 Estimates of Parameters, and Their Standard Errors, Obtained by Applying Step (B) to the Linear Model (7.2)

$\hat{\alpha}_{MG} = .00353925 \pm .00036817$	$\hat{\beta}_{MG} = .000113563 \pm .000004887$
$\hat{\alpha}_{U} = .00384017 \pm .00035198$	$\hat{\beta}_{U} = .000119745 \pm .000004683$
$\hat{\alpha}_{PL} = .00184907 \pm .00036899$	$\hat{\beta}_{PL} = .0000925069 \pm .000004128$
$\hat{\alpha}_{V} = .00200926 \pm .00035522$	$\hat{\beta}_{V} = .000137906 \pm .000004776$

Brown Imperial Spanish suggest that this parameter is invariant for this variety at the two sites MG and U. For White Imperial Spanish, however, the values of β seem different. This will be thoroughly investigated in Sec. 7.3.

7.3 Comparing Parameter Estimates in Nonlinear Models

The approach to be used here follows that advocated by Mead (1970) with some minor modifications. As recognized by Mead (1970), it is only rarely that interest in a family of response curves will be limited to fitting the curve for each individual set of data. The more usual circumstance is that one will be interested in comparing the curves for different sets of data, and this means, in effect, testing for invariance of some or all of the parameters. As was the case with the linear version of the yield-density models studied in Sec. 7.2, the procedures for the nonlinear model are best illustrated by example. We utilize the data sets of Appendix 3.A as before, and study the Holliday yield-density model with the assumption that log Y has constant variance; that is, we examine the following model:

$$E(\log Y) = -\log (\alpha + \beta X + \gamma X^2) \qquad (7.8)$$

The first step is to decide whether the term γX^2 in (7.8) is necessary. A formal testing procedure will be advocated which parallels that used in Sec. 7.2, except that in addition to the steps followed there, we will proceed in an even more logical fashion by preceding the test of whether $\gamma = 0$ by a test of whether there is a common gamma. The following three-step prescription may be used:

(A) Fit α, β, and γ to each individual set of data. The residual sum of squares from each of the four data sets, assuming homogeneity of variance, are added together to produce a pooled residual sum of squares. Alternatively, the 12 parameter estimates may be obtained by minimizing

$$S_A = \sum_{i=1}^{4} \sum_{t=1}^{n_i} [\log Y_t + \log (\alpha_{iA} + \beta_{iA} X_t + \gamma_{iA} X_t^2)]^2 \qquad (7.9)$$

where n_i denotes the sample size of the ith data set. For the iterative Gauss–Newton procedure, initial estimates of α_{iA}, β_{iA}, and γ_{iA}, i = 1, 2, 3, 4, are needed. For the α's and β's, the estimates from Table 7.3 may be used, whereas for the γ's, a reasonable initial choice is $\gamma_{iA} = 0$, i = 1, 2, 3, 4.

    (B) Fit a common γ to the four data sets, but fit individual α's and β's. To do this involves minimizing the sum of squares

$$S_B = \sum_{i=1}^{4} \sum_{t=1}^{n_i} [\log Y_t + \log (\alpha_{iB} + \beta_{iB}X_t + \gamma_B X_t^2)]^2 \qquad (7.10)$$

The absence of the subscript i on γ_B indicates that this parameter will have the same estimate for each of the four data sets. The same initial estimates of the parameters that were used in (A) may be used here.

    (C) Fit α and β to each individual set of data. The residual sum of squares from each of the four data sets are added together to produce a pooled residual sum of squares, or alternatively, the eight parameter estimates and pooled residual sum of squares may be obtained by minimizing

$$S_C = \sum_{i=1}^{4} \sum_{t=1}^{n_i} [\log Y_t + \log (\alpha_{iC} + \beta_{iC}X_t)]^2 \qquad (7.11)$$

The estimates of the α's and β's from Table 7.3 may be used as initial estimates.

    The results of following the prescription above are summarized in Table 7.4, in a manner analogous to that used in Table 7.1. The RSS obtained from step (A) (i.e., $RSS_A = 1.71577$) and the RMS of .011141 based on it, serves as a basis for making tests of more restricted assumptions about the values of some parameters. Step (A) corresponds to the most general assumption that one can make about the fitting of model (7.8), namely that different values of α, β, and γ apply to each of the four data sets. The next more restrictive model is that embodied in the assumptions of step (B), where it is postulated that there may be a common γ. The test of a common γ involves examining the difference between the RSS's of steps (B) and (A), obtaining the mean square by

Table 7.4 Comparison of Fits for Onion Data of Appendix 3.A with Parameter γ Included and Excluded, for Nonlinear Model Assuming Var (log Y) Constant

Description of fit or test	Number of parameters estimated, p	Degrees of freedom, df	Residual sum of squares, RSS	Residual mean square, RMS
(C) Individual α and β, with γ = 0	8	158	1.74019	
(B) Individual α and β, common γ	9	157	1.73532	
(A) Individual α, β, and γ	12	154	1.71577	.011141

	df	Change in RSS	Mean square	Variance ratio, F
(B) − (A): test of common γ	3	.01955	.006517	$.58^{ns}$
(C) − (A): test of $\gamma = 0$	4	.02442	.006105	$.55^{ns}$

Note: $^{ns}p > .05$.

dividing the difference by the degrees of freedom, which is the difference between the number of parameters estimated in steps (B) and (A), and then dividing by .011141 to obtain an F-value of .58 (see Table 7.4), which in this case is not significant. Since a common value of γ is acceptable, it is logical to ask whether that common γ might have the value zero, thereby implying that an "asymptotic" yield-density model would be suitable for these data. Step (C) leads to the necessary numerical results for testing the hypothesis $\gamma = 0$, and from Table 7.4 the difference between (C) and (A) gives a sum of squares of .02442 with 4 degrees of freedom, so that the mean square of .006105, when divided by .011141, results in an F-value of .55, which is not significant. Hence one may conclude that the asymptotic yield-density model is acceptable; all subsequent tests of invariance are based on the model

$$E(\log Y) = -\log (\alpha + \beta X) \tag{7.12}$$

We may next proceed in a manner parallel to that of Sec. 7.2 and examine whether there is an invariant value of α and/or β over the four data sets. However, as was the case with the linear version of model (7.12), namely model (7.2), the hypothesis of invariant α's or β's is rejected for these data; the demonstration of this is left to the reader as an exercise (see Exercise 7.2). Nevertheless, as indicated in the latter part of Sec. 7.2, although there is no invariant α or β for all four data sets, there is likely to be an invariant α for each of the two varieties separately, and there is also the possibility of an invariant β for the variety Brown Imperial Spanish. We will now put forward the following four-step prescription for examining these possible invariances. They are to be carried out separately for each of the two varieties, each variety being grown at each of two localities.

(D) Fit α and β to each of the two individual sets of data. This provides the most general, or least restricted, model for carrying out tests of subsequent hypothesis. Each of the two data sets may be fitted individually and their residual sum of squares pooled, or the same result may be obtained by minimizing

$$S_D = \sum_{i=1}^{2} \sum_{t=1}^{n_i} [\log Y_t + \log (\alpha_{iD} + \beta_{iD} X_t)]^2 \qquad (7.13)$$

The estimates of the α's and β's in Table 7.3 may be used as initial estimates.

(E) Fit a common α and β to each of the two data sets. This involves minimizing

$$S_E = \sum_{i=1}^{2} \sum_{t=1}^{n_i} [\log Y_t + \log (\alpha_E + \beta_E X_t)]^2 \qquad (7.14)$$

which is equivalent to fitting model (7.12) to the combined 83 observations within a variety.

(F) Fit a common α to the two data sets, but fit individual β's. This involves minimizing

$$S_F = \sum_{i=1}^{2} \sum_{t=1}^{n_i} [\log Y_t + \log (\alpha_F + \beta_{iF} X_t)]^2 \tag{7.15}$$

(G) Fit a common β to the two data sets, but fit individual α's. This involves minimizing

$$S_G = \sum_{i=1}^{2} \sum_{t=1}^{n_i} [\log Y_t + \log (\alpha_{iG} + \beta_G X_t)]^2 \tag{7.16}$$

The results of carrying out the sequence of fits above are presented in Table 7.5. For Brown Imperial Spanish, the hypothesis of an invariant α and β is only just rejected (F = 3.615, 2 and 154 df, P < .05) after testing for invariance by taking the difference between the RSSs obtained from steps (E) and (D) and dividing by .011141, the RMS obtained from step A of Table 7.4 and based on the pooled RSS under the hypothesis that the variances are equal for all four data sets. Results obtained using the differences (F) − (D) and (G) − (D), respectively, show that an invariant α (ignoring β) and an invariant β (ignoring α) would each be readily accepted. The existence of a common α is not surprising, as α is interpreted (Chapter 3) as a genetic parameter and the data under consideration relate to a single variety. The closeness of the estimates of β, which is interpreted as an environmental parameter, is possibly fortuitous. Mt. Gambier is a coastal city with an average annual rainfall of 712 mm, whereas Uraidla is located in the Adelaide Hills and has an annual rainfall of 1099 mm. Uraidla's colder climate would tend to reduce growth, whereas its higher rainfall would tend to increase growth; the two compensating effects may result in a growth rate for onions that is comparable to that at Mt. Gambier. For White Imperial Spanish, the hypothesis of an invariant α is acceptable (F = .041, 1 and 79 df), but the hypothesis that β is also invariant is strongly rejected (P < .01). The lack of invariance of the environmental parameter β is not surprising, as Purnong Landing is located in a drier climate than is Virginia (average annual rainfall 294 mm compared with 436 mm) and it is to be expected that the yield at Purnong Landing would be less, for the same plant density, than at Virginia.

Table 7.5 Comparison of Fits for Onion Data of Appendix 3.A to Test Invariance of α and/or β for Each Variety Separately, for Nonlinear Model Assuming Var (log Y) Constant

A. Brown Imperial Spanish at Mt. Gambier and Uraidla

Description of fit or test	Number of parameters estimated, p	Degrees of freedom, df	Residual sum of squares, RSS	Residual mean square, RMS
(G) Common β	3	80	.911560	
(F) Common α	3	80	.901143	
(E) Common α and β	2	81	.980853	
(D) Individual α and β	4	79	.900304	.011396

	df	Change in RSS	Mean square	Variance ratio, F^a
(E) − (D): test of invariant α and β	2	.080549	.040275	3.615^*
(F) − (D): test of invariant α	1	.000839	.000839	$.075^{ns}$
(G) − (D): test of invariant β	1	.011256	.011256	1.010^{ns}

B. White Imperial Spanish at Purnong Landing and Virginia

Description of fit or test	Number of parameters estimated, p	Degrees of freedom, df	Residual sum of squares, RSS	Residual mean square, RMS
(G) Common β	3	80	1.30941	
(F) Common α	3	80	.84032	
(E) Common α and β	2	81	2.97878	
(D) Individual α and β	4	79	.83989	.01063

	df	Change in RSS	Mean square	Variance ratio, F^a
(E) − (D): test of invariant α and β	2	2.13889	1.06945	96.0^{**}
(F) − (D): test of invariant α	1	.00043	.00043	$.039^{ns}$
(G) − (D): test of invariant β	1	.46952	.46952	42.1^{**}

Note: $^{ns}p > .05$; $^*p < .05$; $^{**}p < .01$.

[a] The denominator for calculating F is obtained from line (A) of Table 7.4.

The estimated parameters and their associated standard errors, obtained as a by-product of the estimation procedure, are presented in Table 7.6 for both onion varieties for the model of invariant α, the model leading to expression (7.15). The parameter estimates in Table 7.6 compare well with the estimates in Table 7.3 obtained for the linear form of the asymptotic yield-density relationship. This suggests that, if these data are typical, the difference between assuming Var (log Y) constant and assuming Var (1/Y) constant is unimportant.

7.4 Discussion of Comparison of Parameters Procedures

Sections 7.2 and 7.3 illustrated how the examination of the difference between the residual sum of squares for the model making the least restrictive assumption about the parameters and that for other models with more restrictive assumptions about the parameters could be used to make decisions about parameter invariance. Section 7.2 was devoted to linear models which pose a simpler problem than nonlinear models for the following reasons:

1. The intrinsic nonlinearity (Sec. 2.4) of any linear model is zero and therefore there can be no bias in the difference between the residual sum of squares used to draw inferences about parameter invariance. Therefore, tests of significance between the model incorporating the least restrictive parameter assumptions and models with more restrictive parameter assumptions using the

Table 7.6 Parameter Estimates, and Their Standard Errors, for Model with Invariant α, with Var (log Y) Assumed Constant

Brown Imperial Spanish	White Imperial Spanish
$\hat{\alpha}$ = .00380407 ± .000243156	$\hat{\alpha}$ = .00191301 ± .000179058
$\hat{\beta}_{MG}$ = .000108275 ± .000004833	$\hat{\beta}_{PL}$ = .000091148 ± .000003230
$\hat{\beta}_{U}$ = .000119400 ± .000005125	$\hat{\beta}_{V}$ = .000138887 ± .000004328

"extra sum of squares" principle are statistically exact (see Milliken and DeBruin, 1978).

2. The LS estimates of the parameters of a linear model, such as those corresponding to the steps of the prescription detailed in Sec. 7.2, can always be obtained readily. For example, use of the Gauss-Newton algorithm will produce the correct LS estimates in one iteration from any set of initial values.

Therefore, the user will encounter no difficulties in implementing the suggested test procedure for linear models. The situation is vastly different for the nonlinear models of Sec. 7.3 which pose difficulties for the following reasons:

1. Nonlinear models exhibit, in general, an intrinsic nonlinearity, the degree of which varies greatly from model to model, as may be seen from the examples of Chapters 3-6. As inferences about parameter invariance are based on differences between the residual sum of squares of two models, bias is likely, owing to the different extents of intrinsic nonlinearity in the two models being compared. Results obtained using the "extra sum of squares" principle are only approximate, becoming exact only asymptotically (Milliken and DeBruin, 1978). The extent of the intrinsic nonlinearity determines the goodness of the approximation.

2. The LS estimates of the parameters of nonlinear models, such as those corresponding to the steps of the prescription in Sec. 7.3, may not be obtained readily, and may depend on the user being able to supply good initial estimates of the parameters. Unless the LS estimates can be obtained, the tests of invariance cannot be performed.

The above two reasons provide strong justification for the modeler to attempt to find a model that exhibits close-to-linear behavior, *both* with respect to the intrinsic component of nonlinearity and the parameter-effects component of nonlinearity. The intrinsic component of nonlinearity relates to the first of the two points above, as the more closely the solution locus of a model approaches linearity, the less is the expected bias in the predicted Y-values and in the residual sum of squares. The parameter-effects component of nonlinearity relates to the second of the points above, as the greater is this component, the greater are the difficulties in obtaining the LS estimates. Reparameterization to produce a model function with smaller parameter-effects nonlinearity, provided that the intrinsic nonlinearity is also small, is therefore a desirable goal for the modeler to strive for, as it increases the chances of obtaining the LS estimates readily.

It is particularly important that the modeler remain aware of the fact that the prescriptions given in Secs. 7.2 and 7.3 are *hierarchical* ones, meaning that inferential decisions made at a later stage of the decision-making process depend on the decisions made at the earlier stages. Under these circumstances it is well known that the nominal significance levels for tests in the later stages may be quite different from the true, albeit unknown, significance levels, even for those tests where one might expect the significance levels to be exact, as, for example, in the linear models of Sec. 7.2. The problem is that later-stage tests which depend on early-stage decisions involve conditional probability distributions. The modeler, therefore, has little recourse other than to adopt a commonsense approach to model building and decision making in such questions as parameter invariance. In these matters, a significance level can only be useful as a guideline. This is especially true for nonlinear regression models, where the nominal significance levels may be grossly different from the true significance levels if the degree of intrinsic nonlinearity is high.

Exercises

7.1 In Table 7.1, a test was made of the hypothesis $\gamma = 0$ in model (7.1) as applied to the four sets of data on onions from Appendix 3.A. Instead, fit a similar model but with a common $\gamma \neq 0$ to the four data sets, while allowing individual α's and β's. Why must the residual sum of squares for this model lie between RSS_A and RSS_B?

7.2   Show that if all four data sets on onions in Appendix 3.A are simultaneously considered, omitting the suspected outliers, neither parameter α nor parameter β in model (7.12) is invariant.

7.3   Table 7.3 presents LS parameter estimates and their standard errors for α and β for the linear yield-density model (7.2) for each of the four sets of data on onions in Appendix 3.A. Within each variety, at the two sites for which each variety was grown, it is possible to test for parameter invariance using the t-test

$$t = \frac{\hat{\theta}_1 - \hat{\theta}_2}{\sqrt{2\{(df_1)[Var (\hat{\theta}_1)] + (df_2)[Var (\hat{\theta})]\}/(df_1 + df_2)}}$$

where $Var (\hat{\theta}_i) = [SE (\hat{\theta}_i)]^2$, with $SE (\hat{\theta}_i)$ being the standard error of the LS estimate of θ_i at the ith site, $i = 1, 2$, and df_1 and df_2 being the degrees of freedom associated with the residual variance at the first and second sites, respectively. The degrees of freedom associated with t is $df_1 + df_2$. Apply this formula to Brown Imperial Spanish to test for invariance of (a) α and (b) β, respectively, and to White Imperial Spanish to test for invariance of (c) α and (d) β, respectively.

7.4   The inference using the formula in Exercise 7.3 will be exactly the same at that which would result from the F-test derived using the "extra sum of squares" principle for the linear yield-density model (7.2). This follows from the fact that $F_{1,df} = t_{df}^2$. Carry out calculations based on the extra sum of squares principle for model (7.2) for the data in Appendix 3.A for the onion varieties White Imperial Spanish and Brown Imperial Spanish separately and confirm the exactness of this equivalence.

7.5   Consider the following yield-density data for the yield of sweet lupins (*Lupinus angustifolius* cv. Unicrop) grown at the Elliott

Research Farm of the Tasmanian Department of Agriculture for each
of two growing seasons:

Density, X (plants/m^2)	Yield per plant, Y (g) 1975–1976	1976–1977
400	1.2	1.0
100	3.9	3.5
49	8.6	7.1
25	13.6	12.4
11	25.7	22.8
4	54.3	37.6

(a) For model (7.2), test for the invariance of parameter α over
     the two growing seasons. If an invariant α is accepted, test
     for the invariance of β.

(b) Show that the results are identical irrespective of whether
     the tests for an invariant α or β are carried out using
     F-tests based on the extra sum of squares principle, or
     carried out using t-tests employing the LS estimates of the
     parameters and their pooled variances; that is, demonstrate
     that $F_{1,df} = t_{df}^2$ for each test, where df is the appropriate
     degrees of freedom. Demonstrate that this is not the case,
     however, for the test of an invariant β, given that there
     is a common α.

(c) Repeat the tests in (a) assuming instead that Var (log Y)
     is constant; that is, consider model (7.12) instead of (7.2).

(d) Show that the F-test for the invariance of α or β via the
     extra sum of squares principle is *not* equivalent to the
     t-test approach for the tests in (c); that is, demonstrate
     that $F_{1,df} \neq t_{df}^2$, where df is the appropriate degrees of
     freedom, for this nonlinear model.

7.6 Consider the following data on potato yields Y versus rate of
     fertilizer application X obtained by Pimentel-Gomes (1953) for
     each of two growing seasons:

X	Y	
	1945	1946
0	232.65	104.75
1	369.08	188.63
2	455.63	211.75
3	491.45	217.63
4	511.50	231.13

Using model function (5.1), carry out tests of parameter invariance for each of the three parameters α, β, and γ.

8

Obtaining Good Initial Parameter Estimates

8.1 Introduction

The method advocated in this book for obtaining the LS estimates of the parameters in nonlinear models is the Gauss-Newton method; a program for its use appears in the Appendix to this book. This method will always converge to the LS estimates provided that the initial parameter estimates are sufficiently close to the LS estimates. Numerous other methods have been proposed for obtaining the LS estimates, many of which involve modification of the Gauss-Newton method, by either altering the size of the correction vector for the estimates at each step in the iterative procedure, or by altering the direction of search. Some of these methods are described in detail by Bard (1974) and Kennedy and Gentle (1980), to which interested readers are referred. However, as stressed by Kennedy and Gentle (1980), these methods do not necessarily work in the absence of good initial estimates, which appear to be essential to guarantee convergence of the algorithm employed. With good initial estimates, the unmodified Gauss-Newton procedure will also converge, rendering the modifications unnecessary. Good initial estimates are required in a simulation study, where it is absolutely essential to obtain the converged LS estimates for each one of the trials. Unconverged data sets cannot be discarded and replaced by additional data sets, as this must lead to biased results. For many of the models studied in Chapters 3-6, convergence in simulation studies was achieved by using as initial parameter estimates the assumed true parameter values that were employed to generate the data. For badly behaved nonlinear models, however, this was not adequate. Therefore, this chapter is devoted to formal procedures for obtaining good initial estimates of the parameters. Examples will be presented for all models studied in Chapters 3-5, and it will be left to the reader as exercises to derive good initial estimates for the models of Chapter 6.

8.2   Initial Parameter Estimates for Yield-Density Models

In this section we derive formulas which lead to good initial estimates of parameters for the Holliday, Bleasdale-Nelder, and Farazdaghi-Harris models studied in Chapter 3. As determined in Chapter 3, the Holliday model behaves like a linear model, and therefore convergence to the LS estimates will usually be achieved for this model even from rather crude initial estimates. Nevertheless, formulas will be presented for obtaining good initial parameter estimates for this model as well as for the other models, as the information or insight gained may prove useful to the modeler when other models have to be considered.

8.2.1   Holliday model

The deterministic component of the Holliday model is given by (3.2) as

$$Y = (\alpha + \beta X + \gamma X^2)^{-1}$$

For the additive error assumption, with Var (Y) assumed constant throughout the range of the data, the formula above becomes

$$E(Y) = \frac{1}{\alpha + \beta X + \gamma X^2} \tag{8.1}$$

and for the multiplicative error assumption, with Var $(\log Y)$ assumed constant throughout the range of the data, one obtains

$$E(\log Y) = -\log (\alpha + \beta X + \gamma X^2) \tag{8.2}$$

A third error assumption can be made which results in

$$E\left(\frac{1}{Y}\right) = \alpha + \beta X + \gamma X^2 \qquad (8.3)$$

the assumption being that the variance of the reciprocal of Y is constant throughout the range of the data. Model (8.3) is clearly linear in the parameters α, β, and γ, and therefore the least-squares problem of minimizing

$$S = \sum_{t=1}^{n} \left[\frac{1}{Y_t} - (\alpha + \beta X_t + \gamma X_t^2)\right]^2 \qquad (8.4)$$

where the summation is carried out over the n data observations, results in the following set of linear equations whose solution is easily obtained by standard procedures for linear regression:

$$n\alpha + \beta \sum X + \gamma \sum X^2 = \sum \left(\frac{1}{Y}\right)$$

$$\alpha \sum X + \beta \sum X^2 + \gamma \sum X^3 = \sum \left(\frac{X}{Y}\right) \qquad (8.5)$$

$$\alpha \sum X^2 + \beta \sum X^3 + \gamma \sum X^4 = \sum \left(\frac{X^2}{Y}\right)$$

In the notation above it is assumed that readers understand that the summations are to be carried out over the n data observations, the subscripts on the X's and Y's having been dropped to simplify the notation. For example,

$$\sum \left(\frac{X}{Y}\right) \quad \text{signifies} \quad \sum_{t=1}^{n} \frac{X_t}{Y_t}$$

The Gauss-Newton program in the Appendix to this book will find the LS estimates in a single iteration from any starting vector of initial estimates. For example, the null vector (α = 0, β = 0, γ = 0) will suffice as an initial estimate. Although the procedure above is based on the assumption that Var (1/Y) is constant, which is an entirely different assumption about the stochastic term than either the additive or multiplicative error assumption, the estimates, denoted α_0, β_0, and γ_0, respectively, obtained by solving the system of linear equations (8.5), should be adequate initial estimates to lead to convergence when the Gauss-Newton algorithm is used with either model (8.1) or (8.2).

We now examine a somewhat more accurate approximate procedure for obtaining initial estimates for the multiplicative error assumption based on a weighted least-squares criterion that also leads to a set of linear equations. The weighted least-squares criterion is the same as that used by Mead (1970), and the procedure to be presented here closely follows his treatment. A difference, however, is that Mead (1970) often used weighted least-squares as an approximation to the exact least-squares approach, whereas in the present treatment we use it to obtain initial estimates for subsequent use with an exact procedure. The weighted least-squares criterion is based on minimizing the expression

$$S = \sum_{t=1}^{n} \frac{\sigma^2}{Var\ (Y_t)} \left[Y_t - E(Y_t) \right]^2 \tag{8.6}$$

If $Var\ (Y_t)$ is constant (say σ^2) throughout the range of the data, as it is for the additive error assumption, expression (8.6) reduces to the usual LS criterion.   However, if multiplicative rather than additive error is assumed, then,

$$Var\ (\log Y_t) = \sigma^2 \quad \text{for all } t, \ t = 1, \ 2, \ \ldots, \ n$$

The variance of the Y's [i.e., $Var\ (Y_t)$], which must be substituted into (8.6), can be approximately obtained in terms of the variance of some function of Y_t (such as $\log Y_t$ as above) by means of the following formula (see, e.g., Kendall and Stuart, 1963, Sec. 10.7),

$$Var\ [f(Y_t)] \approx \left[\frac{\partial f(Y_t)}{\partial Y_t} \right]^2_{Y_t=E(Y_t)} Var\ (Y_t) \tag{8.7}$$

where the subscript $Y_t = E(Y_t)$ means that the derivative is to be evaluated at $E(Y_t)$. Rearranging (8.7), one obtains

$$Var (Y_t) \approx \frac{Var [f(Y_t)]}{[\partial f(Y_t)/\partial Y_t]^2_{Y_t=E(Y_t)}} \tag{8.8}$$

For $f(Y_t) = \log Y_t$, expression (8.8) gives

$$Var (Y_t) = \frac{Var (\log Y_t)}{(1/Y_t)^2_{Y_t=E(Y_t)}}$$

so that

$$Var (Y_t) = \sigma^2 E^2(Y_t) \tag{8.9}$$

Substituting (8.9) into (8.6), the following is obtained:

$$S = \sum_{t=1}^{n} \left[\frac{Y_t}{E(Y_t)} - 1 \right]^2 \tag{8.10}$$

Since $E(Y_t) = 1/(\alpha + \beta X_t + \gamma X_t^2)$ for the Holliday model [cf. expression (8.1)], one obtains, after substitution into (8.10), the subscripts being dropped to simplify the notation:

$$S = \sum [Y(\alpha + \beta X + \gamma X^2) - 1]^2 \tag{8.11}$$

The "normal" equations for solving for the parameter estimates are obtained by differentiating the right hand side of (8.11) with respect to α, β, and γ successively, and setting each resulting expression equal to zero. This yields the set of equations

$$\alpha \sum Y^2 + \beta \sum XY^2 + \gamma \sum X^2Y^2 = \sum Y$$
$$\alpha \sum XY^2 + \beta \sum X^2Y^2 + \gamma \sum X^3Y^2 = \sum XY \tag{8.12}$$
$$\alpha \sum X^2Y^2 + \beta \sum X^3Y^2 + \gamma \sum X^4Y^2 = \sum X^2Y$$

The solution of the system of linear equations (8.12) gives estimates, denoted here α_0, β_0, and γ_0, respectively, which may be employed with the Gauss-Newton algorithm to obtain the exact LS solution based on model (8.2). As an exercise, the reader may show, using straightforward but tedious algebra, that the residual sum of squares obtained using these estimates is

$$RSS = n - \alpha_0 \sum Y - \beta_0 \sum XY - \gamma_0 \sum X^2 Y \qquad (8.13)$$

8.2.2 Farazdaghi-Harris model

The Farazdaghi—Harris model, for the additive error assumption, is

$$E(Y) = \frac{1}{\alpha + \beta X^\phi} \qquad (8.14)$$

and, for the multiplicative error assumption, is

$$E(\log Y) = -\log (\alpha + \beta X^\phi) \qquad (8.15)$$

By analogy with a method employed in Section 8.2.1, we may also consider the error assumption of constant variance for the reciprocal of Y in (3.3), which leads to

$$E\left(\frac{1}{Y}\right) = \alpha + \beta X^\phi \qquad (8.16)$$

The parameter ϕ lies between 0 and 1, and for a fixed value of ϕ, model (8.16) is linear in the remaining parameters and may therefore be fitted by ordinary linear least squares. An adequate procedure might be to start with $\phi_0 = .05$ (say) and solve for the values of α and β, denoted α_0 and β_0, respectively, which minimize

$$S = \sum_{t=1}^{n} \left(\frac{1}{Y_t} - \alpha - \beta X_t^{\phi_0}\right)^2 \qquad (8.17)$$

by linear least squares. Then ϕ_0 may be incremented in steps of .05 (say) and α_0 and β_0 redetermined by minimizing S in (8.17) at each step, and denoting the residual sum of squares by RSS. The residual sum of squares RSS will decrease monotonically as ϕ_0 is increased until ϕ_0 is near its optimum value, after which RSS will increase. The set of estimates α_0, β_0, and ϕ_0 which gives the smallest observed RSS should provide good initial estimates for the Gauss—Newton procedure. Greater accuracy can be obtained by using a smaller step size for incrementing ϕ_0. Although the use of expression (8.17) minimizes a

different model, namely (8.16), from either (8.14) which is associated
with the additive error assumption or (8.15) for the multiplicative
error assumption, the estimates α_0, β_0, and ϕ_0 obtained for (8.16) may
serve as adequate starting values for either of the other two error
assumptions.   However, for the multiplicative error assumption
embodied in model (8.15), greater accuracy can be obtained by using
the weighted least-squares criterion.   By following the same
development as in Sec. 8.2.1, the following expression is obtained,
which is identical with (8.10):

$$S = \sum_{t=1}^{n} \left[\frac{Y_t}{E(Y_t)} - 1 \right]^2$$

Since $E(Y_t) = 1/(\alpha + \beta X_t^\phi)$ [cf. expression (8.14)], substitution into
(8.10) leads to the following expression, after dropping subscripts:

$$S = \sum \left[Y(\alpha + \beta X^\phi) - 1 \right]^2 \tag{8.18}$$

For a fixed value of ϕ, denoted ϕ_0, the "normal" equations are

$$\alpha \sum Y^2 + \beta \sum X^{\phi_0} Y^2 = \sum Y$$
$$\alpha \sum X^{\phi_0} Y^2 + \beta \sum X^{2\phi_0} Y^2 = \sum X^{\phi_0} Y \tag{8.19}$$

Solution of the system of linear equations (8.19) gives α_0 and β_0, which
may be substituted back into expression (8.18) to obtain the residual
sum of squares RSS corresponding to the trial value of ϕ_0.   Instead
of substituting back into (8.18), RSS may also be obtained from

$$RSS = n - \alpha_0 \sum Y - \beta_0 \sum X^{\phi_0} Y \tag{8.20}$$

The trial values of ϕ_0 may be incremented in steps of .025 or .05 (say)
from starting values of .025 or .05 (say) and the set of estimates α_0,
β_0, and ϕ_0 which results in the smallest value of RSS may be used as
initial estimates in the exact Gauss-Newton procedure to obtain the
LS estimates for model (8.15), which embodies the multiplicative error
assumption.   Also, α_0, β_0, and γ_0 are suitable initial estimates for

the exact Gauss–Newton procedure based on model (8.14), which embodies the additive error assumption. This procedure was used to obtain the results for the simulation studies reported in Chapter 3 for the Farazdaghi–Harris model.

8.2.3 Bleasdale-Nelder model

The Bleasdale–Nelder model for the additive error assumption is

$$E(Y) = (\alpha + \beta X)^{-1/\theta} \tag{8.21}$$

and, for the multiplicative error assumption, is

$$E(\log Y) = -\frac{1}{\theta} \log (\alpha + \beta X) \tag{8.22}$$

For the latter assumption, good initial estimates may be obtained using the approach employed by Mead (1970). To use this approach, model (3.1) is rewritten as

$$Y^\theta = \frac{1}{\alpha + \beta X} \tag{8.23}$$

The crux of the procedure is now to write all expressions in terms of Y^θ instead of Y. For example, the weighted least–squares criterion (8.6) is rewritten as

$$S = \sum_{t=1}^{n} \frac{\sigma^2}{Var\ (Y_t^\theta)} \ [Y_t^\theta - E(Y_t^\theta)]^2 \tag{8.24}$$

For the multiplicative error assumption, $Var\ (\log Y_t) = \sigma^2$. From expression (8.8), it follows that

$$Var\ (Y_t^\theta) \approx \frac{Var\ (\log Y_t)}{(\partial \log Y_t / \partial Y_t^\theta)^2_{Y_t^\theta = E(Y_t^\theta)}}$$

$$= \sigma^2 \Bigg/ \left(\frac{\partial \log Y_t}{\partial Y_t} \ \frac{\partial Y_t}{\partial Y_t^\theta} \right)^2_{Y_t^\theta = E(Y_t^\theta)}$$

$$= \sigma^2 (\theta Y_t^{\theta-1} \ Y_t)^2_{Y_t^\theta = E(Y_t^\theta)}$$

$$= \sigma^2 \theta^2 E^2 (Y_t^\theta) \tag{8.25}$$

Substituting (8.25) and (8.23) into (8.24) yields, after dropping subscripts for simplicity,

$$S = \theta^{-2} \sum [Y^\theta (\alpha + \beta X) - 1]^2 \tag{8.26}$$

For a fixed value of θ, denoted θ_0, the "normal" equations are

$$\alpha \sum Y^{2\theta_0} + \beta \sum XY^{2\theta_0} = \sum Y^{\theta_0}$$
$$\alpha \sum XY^{2\theta_0} + \beta \sum X^2 Y^{2\theta_0} = \sum XY^{\theta_0} \tag{8.27}$$

Solution of the system of linear equations (8.27) gives α_0 and β_0, which may be substituted back into expression (8.26) to obtain the residual sum of squares RSS corresponding to the trial value of θ_0. Alternatively, by algebraic manipulations of (8.26) and (8.27), it may be shown that RSS is given by

$$RSS = \theta^{-2} \left(n - \alpha_0 \sum Y^{\theta_0} - \beta_0 \sum XY^{\theta_0} \right) \tag{8.28}$$

The trial values of θ_0 may be incremented in steps of .025 or .05 (say) from starting values of .025 or .05 (say) and the set of estimates α_0, β_0, and θ_0 which results in the smallest value of RSS may be used as an initial estimate in the exact Gauss–Newton procedure to obtain the LS estimates for model (8.22), which embodies the multiplicative error assumption. It is possible that α_0, β_0, and θ_0 may also be suitable initial estimates for the exact Gauss–Newton procedure based on model (8.21), which embodies the additive error assumption. This procedure was used to obtain the results for the simulation studies reported in Chapter 3 for the Bleasdale–Nelder model.

8.3 Initial Parameter Estimates for Sigmoidal Growth Models

In this section we derive or present formulae which lead to good initial estimates of the parameters for the five sigmoidal growth models studied in Chapter 4. As is to be expected, there is more than one

way to obtain initial parameter estimates, and therefore more than one
method will be presented for some models. These do not exhaust the
possibilities available for obtaining initial estimates of the
parameters of these models, but they will illustrate to the reader a
range of methods that may be employed. It will be difficult to judge
which of two or more of these methods is "best"; this does not matter
as long as the goal is achieved of finding initial estimates which,
when they are then subsequently used in conjunction with a procedure
such as Gauss-Newton, will produce the LS estimates in only a few
iterations.

8.3.1 Gompertz model

Consider the model function of the Gompertz model given by expression
(4.1), namely

$$Y = \alpha \exp [-\exp (\beta - \gamma X)]$$

The first step in obtaining initial estimates, which is equally valid
for the other models to be studied in this section as well, is to draw
a graph of the data in the form Y versus X. Such a graph is given in
Fig. 4.1 for the four data sets on vegetative growth processes; the
solid line, which represents the fitted logistic model, should be
ignored and the reader's attention directed only to the data points.
A reasonable initial estimate, denoted α_0, may be obtained visually
for the asymptote in (4.1) as the approximate maximum value approached
by the response Y as $X \to \infty$. Then, one rearranges (4.1) to give

$$Z_0 = \log \left[-\log \left(\frac{Y}{\alpha_0} \right) \right] = \beta - \gamma X \qquad (8.29)$$

Expression (8.29) is a linear model and estimates of β and γ, denoted
β_0 and γ_0, respectively, are obtained from the simple linear regression
of Z_0 on X. The set of estimates α_0, β_0, and γ_0 may then be used as
initial estimates to obtain the LS estimates of the parameters for
either the additive error or the multiplicative error assumptions,
using the Gauss-Newton algorithm.

If convergence using Gauss-Newton is not achieved, however, the user will have to attempt to obtain an estimate α_0 of α more closely than by the freehand procedure suggested above. A range of trial values of α_0 may be used, and for each trial value expression (8.29) would be applied; that is, one would regress Z_0, which is a function of α_0, on X to obtain the estimates β_0 and γ_0 corresponding to the trial value of α_0. The set of estimates α_0, β_0, and γ_0 which results in the smallest value of the residual sum of squares RSS, given by

$$RSS = \sum_{t=1}^{n} [Y_t - E(Y_t)]^2 = \sum_{t=1}^{n} \{Y_t - \alpha_0 \exp [-\exp (\beta_0 - \gamma_0 X_t)]\}^2$$

$$(8.30)$$

may be sufficiently close to the LS estimates $\hat{\alpha}$, $\hat{\beta}$, and $\hat{\gamma}$ so that convergence by the Gauss-Newton algorithm will occur in a few iterations. Use of expression (8.30) is predicated on the assumption of additive error for the stochastic term. If the multiplicative error assumption is made instead, the corresponding formula to use for RSS is

$$RSS = \sum_{t=1}^{n} [\log Y_t - E(\log Y_t)]^2$$

$$= \sum_{t=1}^{n} (\log Y_t - \log \{\alpha_0 \exp [-\exp (\beta_0 - \gamma_0 X_t)]\})^2 \quad (8.31)$$

Consider now the reparameterized Gompertz model given by model function (4.6):

$$Y = \exp (\alpha - \beta \gamma^X)$$

One may obtain the initial estimates α_0, β_0, and γ_0 of the parameters of model function (4.1) by any of the methods detailed above, and simply transform them via the mathematical functions relating the parameters of model function (4.6) with those of model function (4.1). For example, denoting α_{new}, β_{new}, and γ_{new} to be the initial estimates of the parameters α, β, and γ in (4.6), they are related to the initial estimates α_0, β_0, and γ_0 in (4.1) by the formulas

$$\alpha_{new} = \log \alpha_0$$
$$\beta_{new} = \exp(\beta_0) \quad\quad\quad\quad (8.32)$$
$$\gamma_{new} = \exp(-\gamma_0)$$

Alternatively, one may obtain initial estimates of the parameters of (4.6) by rewriting the model function as

$$Z_0 = \log[-(\log Y - \alpha_0)] = \log \beta + X \log \gamma \quad\quad (8.33)$$

where α_0 may be estimated visually from a graph of Y versus X, as was done earlier in this subsection, or α_0 may be one of a series of trial values of α. The expression (8.33) is seen to be linear in $\log \beta$ and $\log \gamma$, and by regressing Z_0 on X, estimates of β and γ, denoted β_0 and γ_0, respectively, may be obtained corresponding to a particular initial estimate α_0.

Another procedure may be employed if the stochastic term can be assumed to be multiplicative. For this error assumption, we may write the reparameterized form corresponding to model (4.6) as

$$E(\log Y) = \alpha - \beta\gamma^X \quad\quad\quad\quad (8.34)$$

Clearly, this model is linear in α and β, but not in γ. However, as γ ranges between zero and unity, it is possible, as was the case with the yield-density models of Sec. 8.2, to derive a systematic procedure for obtaining good initial parameter estimates. An adequate procedure might be to start with an estimate of γ, denoted γ_0, equal to .05 (say), and solve the system of linear least-squares equations resulting from (8.34), namely

$$n\alpha - \beta \sum \gamma_0^X = \sum \log Y$$
$$\alpha \sum \gamma_0^X - \beta \sum \gamma_0^{2X} = \sum \gamma_0^X \log Y \quad\quad (8.35)$$

The parameter estimates obtained from the solution to the system of equations above, denoted α_0 and β_0, together with the trial value γ_0, may then be used to evaluate the residual sum of squares RSS given by

$$RSS = \sum_{t=1}^{n} (\log Y_t - \alpha_0 + \beta_0 \gamma_0^{X_t})^2 \tag{8.36}$$

Subsequently, γ_0 is incremented in steps of .05 (say), and RSS redetermined after solving the system of equations (8.35) at each step. Provided that the initial trial value of γ_0 is sufficiently small, the sum of squares RSS will decrease monotonically as γ_0 is increased until γ_0 is near its optimum value after which RSS will increase. The set of estimates α_0, β_0, and γ_0 corresponding to the smallest RSS observed should provide good initial estimates for the Gauss–Newton procedure.

8.3.2 Logistic model

As we have seen in Chapter 4, the logistic model, in several of its parameterizations studied there, exhibits close-to-linear behavior in estimation. Thus there will usually be little trouble in obtaining convergence when attempting to determine the LS estimates of the parameters of the logistic model from any reasonable set of initial parameter estimates. Consider model function (4.2),

$$Y = \frac{\alpha}{1 + \exp (\beta - \gamma X)}$$

for which the LS estimator exhibits close-to-linear behavior. Often it is sufficient, as was the case with the Gompertz model, to plot the data in the form Y versus X and obtain a visual estimate of the asymptote, denoted α_0. Rearranging, one obtains

$$Z_0 = \log \left(\frac{\alpha_0}{Y} - 1 \right) = \beta - \gamma X \tag{8.37}$$

which is linear in the parameters β and γ, and estimates of these parameters, denoted β_0 and γ_0, respectively, may readily be obtained from the simple linear regression of Z_0 on X. The resulting set of estimates α_0, β_0, and γ_0 may suffice as initial estimates for use with the Gauss–Newton procedure. If not, one can try a range of values of α_0, in a similar fashion to the procedure used for the Gompertz model in Sec. 8.3.1, and by repeatedly applying the linear least-squares expression (8.37) for these different values of α_0, until a sufficiently small RSS is obtained, given by expressions similar to (8.30) and (8.31), depending upon whether additive or multiplicative error has been assumed. This set of α_0, β_0, and γ_0 corresponding to the smallest RSS observed should suffice as initial estimates for use with the Gauss-Newton algorithm.

A somewhat more formal procedure may be employed, especially for use in simulation studies, where it is essential that convergence be achieved for each of the generated data sets. This formal procedure is based on model function (4.8),

$$Y = \frac{1}{\alpha + \beta\gamma^X}$$

and on the weighted least-squares criterion given by expression (8.6). Recall that if a multiplicative error assumption is made, expression (8.6) reduces to (8.10), which, if we drop the subscripts for simplicity, is

$$S = \sum \left[\frac{Y}{E(Y)} - 1 \right]^2$$

Substituting (4.8) into the formula above yields

$$S = \sum [Y(\alpha + \beta\gamma^X) - 1]^2 \tag{8.38}$$

Since γ ranges between zero and unity, an adequate procedure might be to start with an estimate of γ, denoted γ_0, equal to .05 (say), and solve the system of linear equations resulting from (8.38), namely

$$\alpha \sum Y^2 + \beta \sum \gamma_0^X Y^2 = \sum Y$$

$$\alpha \sum \gamma_0^X Y^2 + \beta \sum \gamma_0^{2X} Y^2 = \sum \gamma_0^X Y \tag{8.39}$$

Parameter estimates, denoted α_0 and β_0, obtained from the solution to the system of equations above, together with the trial value γ_0, may then be used to evaluate the residual sum of squares, either by direct substitution into (8.38), or, by algebraic manipulation of (8.38) and (8.39), with α, β and γ replaced by α_0, β_0, and γ_0 to give the expression

$$\text{RSS} = n - \alpha_0 \sum Y - \beta_0 \sum \gamma_0^X Y \qquad (8.40)$$

Subsequently, γ_0 is incremented in steps of .05 (say), and RSS redetermined after solving the system of equations (8.39) at each step. The set of estimates α_0, β_0, and γ_0 corresponding to the smallest RSS observed should provide good initial estimates for the Gauss-Newton procedure, especially for the multiplicative error case.

Model function (4.8) was used in the derivation above in preference to the other model functions of the logistic model studied in Chapter 4 because the form of (4.8) is that of a function whose reciprocal is linear in α and β, and leads to a system of linear equations in those two parameters when substituted into expression (8.10). Having obtained α_0, β_0, and γ_0, however, these may be converted by transformation for use with any model function of the logistic model. For example, if the user is interested in model function (4.2), then the estimate of α in (4.2) is simply $1/\alpha_0$, the estimate of β is log (β_0/α_0), and the estimate of γ is $-\log \gamma_0$.

8.3.3 Richards model

Consider the Richards model in model function (4.3),

$$Y = \frac{\alpha}{[1 + \exp (\beta - \gamma X)]^{1/\delta}}$$

As this model may be considered to be an extension of the logistic model, it may be realistic to use the same procedures as used for the logistic model provided δ can be fixed. After rearranging, one obtains

$$Z = \log \left[\left(\frac{\alpha}{Y} \right)^\delta - 1 \right] = \beta - \gamma X \qquad (8.41)$$

where Z is linear in β and γ and is analogous to expression (8.37) for the logistic model except for the inclusion of δ. Estimates of α and δ may be obtained from graphs of the data in the form Y versus X, such as those presented in Fig. 4.1. For example, a visual estimate of the asymptote α, denoted α_0, may be obtained as the maximum value approached

by the response at high values of X. To obtain an estimate δ_0 of δ,
however, we make use of an estimate of the point of inflection, denoted
(X_F, Y_F), following an approach used by Causton (1969). The point of
inflection is the point, for increasing X, where the slope of the curve
of Y versus X ceases to increase and thereafter begins to fall;
mathematically, it is the point where the second derivative d^2Y/dX^2
is zero. Differentiating (4.3) twice with respect to X, setting the
resulting expression equal to zero, solving for X, and denoting it X_F,
one obtains

$$X_F = \frac{\beta - \log \delta}{\gamma} \qquad\qquad (8.42)$$

Substitution of (8.42) into (4.3) results in the following ordinate
of the point of inflection:

$$Y_F = \alpha(1 + \delta)^{-1/\delta} \qquad\qquad (8.43)$$

Thus a reasonable initial estimate of δ_0 may be obtained by solving
(8.43) using estimates α_0 of the asymptote and of the point of
inflection Y_F. For the data sets studied in Chapter 4, points of
inflection are presented in Table 8.1. These points of inflection are
derived from the fitted model (4.3) using the LS estimates of the
parameters given in Table 4.1 for the additive error case. The success
that users may have in making use of the point of inflection may depend
on their ability to make an estimate of (X_F, Y_F) from a graph of the
data. As an exercise, readers should examine the raw data given in
Fig. 4.1 and see how closely they would estimate the point of inflection
from the data. In so doing, the solid curve in Fig. 4.1, representing
the fitted logistic model, should be ignored.

Having obtained estimates α_0 and δ_0 of α and δ, respectively, these
estimates can be substituted into expression (8.41) to give values of
Z_0 corresponding to each pair of observations (X_t, Y_t), $t = 1, \ldots, n$.
The simple regression of Z_0 on X will then yield values of β_0 and γ_0,
which, together with α_0 and δ_0, may form a suitable set of initial
parameter values for use with the Gauss–Newton algorithm.

Table 8.1 Fitted Points of Inflection (X_F, Y_F) for the Data Sets of Appendix 4.A, Richards Model: Additive Error Assumption

Data set	Point of inflection (X_F, Y_F)
Pasture regrowth	(41.55, 38.93)
Onion bulbs plus tops	(6.623, 367.4)
Cucumber cotyledons	(2.322, 3.353)
Bean root cells	(6.704, 11.70)

If there are sufficient data in the region of low X values, it may be feasible to estimate the expected value of the ordinate intercept, denoted Y_{INT}, this being the response value Y at X = 0. It follows from (4.3) that

$$Y_{INT} = \frac{\alpha}{[1 + \exp(\beta)]^{1/\delta}} \tag{8.44}$$

Having obtained an initial estimate α_0 of the asymptote α from the graph of Y versus X, and an estimate δ_0 of δ by solving (8.43) using the estimate of Y_F as described above, one may obtain an estimate of β, denoted β_0, by rearrangement of expression (8.44) as follows:

$$\beta_0 = \log\left[\left(\frac{\alpha_0}{Y_{INT}}\right)^{\delta_0} - 1\right] \tag{8.45}$$

Rearrangement of (8.42) will then provide an estimate of γ, denoted γ_0, as

$$\gamma_0 = \frac{\beta_0 - \log \delta_0}{X_F} \tag{8.46}$$

Thus from three visual observations on the graph of Y versus X: (1) the asymptote α_0, (2) the ordinate intercept Y_{INT}, and (3) the point of inflection (X_F, Y_F), one is able to determine a set of four initial parameter estimates α_0, β_0, γ_0, and δ_0 for use as starting values in the Gauss-Newton procedure. For the additive error assumption, the residual sum of squares is given by

$$RSS = \sum_{t=1}^{n} \{Y_t - \alpha_0[1 + \exp(\beta_0 - \gamma_0 X_t)]^{-1/\delta_0}\}^2 \qquad (8.47)$$

and for the multiplicative error assumption by

$$RSS = \sum_{t=1}^{n} \left\{ \log Y_t - \log \alpha_0 + \frac{1}{\delta_0} \log [1 + \exp(\beta_0 - \gamma_0 X_t)] \right\}^2$$

$$(8.48)$$

If the user is not confident that the initial estimate α_0 has been well determined, a series of trial values may be used. For each trial value of α_0, δ_0 is redetermined using (8.43), then β_0 redetermined using (8.45), and finally γ_0 from (8.46). Depending on the assumption made about the stochastic term, either expression (8.47) or (8.48) is then used to determine RSS for the set of estimates corresponding to the trial value α_0. The set of estimates giving the smallest value of RSS will be the one to use as starting values for the Gauss-Newton algorithm.

8.3.4 Morgan-Mercer-Flodin (MMF) model

Consider the MMF model given by model function (4.4),

$$Y = \frac{\beta\gamma + \alpha X^\delta}{\gamma + X^\delta}$$

As has been suggested in the previous subsections of this section, a useful initial step is to draw a graph of the data in the form Y versus X, such as those presented in Fig. 4.1. From such a graph, an estimate of the asymptote α, denoted α_0, may be determined as the maximum value approached by the response at high values of X. The ordinate intercept β may also be estimated, denoted β_0, as the value of the ordinate when X = 0. Having determined α_0 and β_0, then at the halfway point between the asymptote α_0 and intercept β_0, that is, at

$$Y_{1/2} = \frac{\alpha_0 + \beta_0}{2} \qquad (8.49)$$

the corresponding value of the abscissa may be readily shown to be

$$X_{1/2} = \gamma^{1/\delta} \tag{8.50}$$

An additional relationship among the parameters may be obtained from an estimate of the point of inflection (X_F, Y_F) using the graph of Y versus X, in a manner identical to that employed for the Richards model (Sec. 8.3.3). It may be shown (see Exercise 8.4) that the coordinates of the point of inflection are given by the following formulas:

$$X_F = \left[\frac{\gamma(\delta - 1)}{\delta + 1} \right]^{1/\delta} \tag{8.51}$$

$$Y_F = \frac{\beta(\delta + 1) + \alpha(\delta - 1)}{2\delta} \tag{8.52}$$

Since α_0 and β_0 have previously been estimated, an estimate of δ, denoted δ_0, may be obtained by solving (8.52) after substituting α_0 and β_0 for α and β, respectively. Then, from (8.50) or (8.51), the remaining parameter estimate γ_0 may be obtained.

Recall that the reparameterized MMF model given by model function (4.14) exhibited more close-to-linear behavior than model function (4.4). The two models differ in that γ in (4.4) is replaced by exp (γ) in (4.14). When model function (4.14) is used, an estimate γ_0 of γ may be written in terms of $X_{1/2}$ as

$$\gamma_0 = \delta_0 \log X_{1/2} \tag{8.53}$$

For the additive error assumption, the residual sum of squares for (4.14) is given by

$$RSS = \sum_{t=1}^{n} \left[Y_t - \frac{\beta_0 \exp (\gamma_0) + \alpha_0 X_t^{\delta_0}}{\exp (\gamma_0) + X_t^{\delta_0}} \right]^2 \tag{8.54}$$

where α_0 and β_0 are estimates of the asymptote and ordinate intercept, respectively, δ_0 is obtained from a solution of (8.52), and γ_0 obtained from (8.53) [or from (8.51) after replacing γ by log γ_0 and δ by δ_0]. A rather similar formula, with the logarithm symbol log inserted before

Y_t and after the minus sign following Y_t, applies for the multiplicative error case. Because model function (4.14) appears to exhibit close-to-linear behavior, if the results for the data sets of Chapter 4 are typical, the Gauss–Newton procedure based on (8.54), or the equivalent formula for the multiplicative error case, has a reasonably good chance of success. Should it fail, however, use of a series of trial values of α_0, with subsequent reevaluation of the other parameters using the formula presented above, will enable the user to obtain a set of estimates α_0, β_0, γ_0, and δ_0 giving the smallest observed residual sum of squares, which may be successful as initial estimates for use with the Gauss–Newton algorithm.

For the data sets studied in Chapter 4, values of $X_{1/2}$ and X_F are presented in Table 8.2, these having been derived from the fitted model (4.4) using the LS estimates of the parameters given in Table 4.1 for the additive error case. The success of the procedure described in this subsection may depend on the reader's ability to estimate $X_{1/2}$ and X_F from the *data*, and as an exercise, readers should examine the graphs in Fig. 4.1 and see how closely they would estimate these quantities. Readers will note from the results in Table 8.2 that X_F < $X_{1/2}$ in every case, a fact that can also be deduced from a comparison of (8.50) and (8.51). The larger the value of δ, the more closely $X_{1/2}$ approaches X_F, which also means the more closely $Y_{1/2}$ approaches Y_F.

Table 8.2 $X_{1/2}$ and X_F Values for the Data Sets of Appendix 4.A, Morgan–Mercer–Flodin Model: Additive Error Assumption

Data set	$X_{1/2}$	X_F
Pasture regrowth	45.739	35.219
Onion bulbs plus tops	6.578	5.986
Cucumber cotyledons	2.815	1.991
Bean root cells	6.633	6.015

8.3.5 Weibull-type model

Consider the Weibull-type model given by model function (4.5),

$$E(Y) = \alpha - \beta \exp (-\gamma x^\delta)$$

As has been suggested in the previous subsections of this section, a useful initial step is to draw a graph of the data observations in the form Y versus X, such as those presented in Fig. 4.1. A reasonable initial estimate of the asymptote α, denoted α_0, may be obtained as the estimated approximate maximum value approached by the response Y as $X \to \infty$. At the other end of the data, for low values of X, an estimate of the ordinate intercept Y_{INT} may be estimated corresponding to $X = 0$. From α_0 and Y_{INT}, an estimate of β, denoted β_0, is then obtained as

$$\beta_0 = \alpha_0 - Y_{INT} \tag{8.55}$$

Having obtained α_0 and β_0 by the procedure above, model function (4.5) can now be rearranged to give

$$Z_0 = \log \left[-\log \left(\frac{\alpha_0 - Y}{\beta_0} \right) \right] = \log \gamma + \delta \log X \tag{8.56}$$

which enables estimates of log γ_0 and δ_0 to be obtained by the simple linear regression of Z_0 on log X. Recall that the reparameterized Weibull-type model, given by model function (4.15),

$$Y = \alpha - \beta \exp [-\exp (-\gamma)x^\delta]$$

exhibited more close-to-linear behavior than model function (4.5). To convert the estimate log γ_0 obtained from (8.56), one need merely take the negative of the exponential of log γ_0 (i.e., one uses $-\gamma_0$) to obtain an estimate of the "new" γ in (4.15). Because model function (4.15) appears to exhibit close-to-linear behavior, if the results for the data sets of Chapter 4 are typical, the Gauss–Newton procedure based on (4.15) has a reasonably good chance of converging successfully. Should it fail, however, use of a series of trial values of α_0, with subsequent reevaluation of the other parameters using the formulas presented above, will enable the user to obtain a set of estimates α_0, β_0, γ_0, and δ_0 corresponding to each trial value of α_0 for which the residual sum of squares RSS for the additive model is

$$RSS = \sum_{t=1}^{n} \{Y_t - \alpha_0 + \beta_0 \exp [-\exp (-\gamma_0)X_t^{\delta_0}]\}^2 \qquad (8.57)$$

and that for the multiplicative error model is

$$RSS = \sum_{t=1}^{n} (\log Y_t - \log \{\alpha_0 - \beta_0 \exp [-\exp (-\gamma_0)X_t^{\delta_0}]\})^2 \quad (8.58)$$

In the two formulas above, α_0 is a trial value of α, β_0 is obtained from (8.55), and γ_0 and δ_0 are obtained from the regression of Z_0 on log X in the expression

$$Z_0 = \log \left[-\log \left(\frac{\alpha_0 - Y}{\beta_0} \right) \right] = -\gamma + \delta \log X \qquad (8.59)$$

The set of estimates α_0, β_0, γ_0, and δ_0 which results in the smallest value of RSS observed, calculated using either (8.57) or (8.58), whichever is the more appropriate, may then be used as initial estimates with the Gauss–Newton algorithm.

As an alternative to linearizing the model function, as in expressions (8.56) and (8.59), to obtain estimates of γ_0 and δ_0, additional relationships may be obtained using the point of inflection (X_F, Y_F) estimated from a graph of Y versus X, in a manner directly analogous to that employed in Sec. 8.3.4 for the MMF model. Differentiating the right–hand side of model function (4.5) twice with respect to X, setting the resulting expression equal to zero, solving for X and denoting the solution as X_F, one obtains

$$X_F = \left(\frac{\delta - 1}{\gamma\delta} \right)^{1/\delta} \qquad (8.60)$$

Substituting (8.60) into (4.5) yields

$$Y_F = \alpha - \beta \exp \left[\frac{-(\delta - 1)}{\delta} \right] \tag{8.61}$$

Thus, after initial estimates α_0 and β_0 of α and β, respectively, have been obtained, an initial estimate of δ, denoted δ_0, may be obtained from a solution of (8.61). Having δ_0, an estimate of γ, denoted γ_0, is readily obtained from a rearrangement of (8.60) as

$$\gamma_0 = \frac{\delta_0 X_F^{\delta_0}}{\delta_0 - 1} \tag{8.62}$$

The value of γ_0 above is readily converted by the transformation $-\log \gamma_0$ for use with model function (4.15), that model function being more likely than model function (4.5) to converge to the LS estimates of the parameters when the Gauss-Newton algorithm is employed.

A further relationship may be obtained from the estimated abscissa $X_{1/2}$ corresponding to the response $Y_{1/2}$ halfway between the ordinate intercept and the asymptote. Since the estimated asymptote is α_0 and the estimated ordinate intercept $Y_{INT} = \alpha_0 - \beta_0$, the halfway point is

$$Y_{1/2} = \alpha_0 - \frac{\beta_0}{2} \tag{8.63}$$

which, when substituted into (4.5) yields

$$X_{1/2} = \left(\frac{\log 2}{\gamma} \right)^{1/\delta} \tag{8.64}$$

Rearrangement of (8.64), together with the use of the estimate δ_0 in place of δ, results in an alternative estimate γ_0 of γ as

$$\gamma_0 = \frac{\log 2}{X_{1/2}^{\delta_0}} \tag{8.65}$$

For the data sets studied in Chapter 4, values of $X_{1/2}$ and X_F are presented in Table 8.3, these having been derived from the fitted model (4.5) using the LS estimates of the parameters given in Table 4.1 for the additive error case. Unlike the MMF model, where $X_F < X_{1/2}$ always, X_F for the Weibull-type model can either be less than or greater than

Table 8.3 $X_{1/2}$ and X_F Values for the Data Sets of Appendix 4.A, Weibull-Type Model

Data set	$X_{1/2}$	X_F
Pasture regrowth	41.211	38.220
Onion bulbs plus tops	6.532	6.533
Cucumber cotyledons	2.731	2.090
Bean root cells	6.537	6.515

$X_{1/2}$ depending on whether or not $(\delta - 1)/\delta < \log 2$. For the data set on the growth of onion bulbs plus tops, the LS estimate $\hat{\delta}$ of δ was 3.26156, which meant that $(\hat{\delta} - 1)/\hat{\delta}$ was almost identical with $\log 2$.

8.4 Initial Parameter Estimates for the Asymptotic Regression Model

In this section we focus our attention on the asymptotic regression model studied in Chapter 5. Consider model function (5.1),

$$Y = \alpha - \beta\gamma^X$$

We may adopt a variety of procedures for obtaining initial parameter estimates, some of which have already been employed in Sec. 8.3.

The most straightforward procedure is probably to draw a graph of the data in the form Y versus X, and obtain a visual estimate, denoted α_0, of the asymptote, estimated to be the approximate maximum value approached by the response Y as $X \rightarrow \infty$. Then, rearranging (5.1) and taking logarithms, one obtains

$$Z_0 = \log (\alpha_0 - Y) = \log \beta + X \log \gamma \qquad (8.66)$$

A linear regression of Z_0 on X will produce estimates of $\log \beta$ and $\log \gamma$, from which β_0 and γ_0, obtained as the exponentials of these estimates, may be used as initial estimates with the Gauss-Newton algorithm. If convergence is not obtained, the modeler might consider using a series of trial values of α_0, redetermining β_0 and γ_0 by the linear regression using (8.66), and then calculating the residual sum of squares RSS for the additive error assumption using

$$RSS = \sum_{t=1}^{n} (Y_t - \alpha_0 + \beta_0 \gamma_0^{X_t})^2 \tag{8.67}$$

or for the multiplicative error assumption using

$$RSS = \sum_{t=1}^{n} [\log Y_t - \log (\alpha_0 - \beta_0 \gamma_0^{X_t})]^2 \tag{8.68}$$

The values α_0, β_0, and γ_0 producing the smallest value of RSS should be used as starting values for the Gauss–Newton procedure.

Alternatively, a series of trial values may be based on the parameter γ. This parameter, in model function (5.1), ranges between zero and unity, which makes it possible to derive a systematic procedure for obtaining good initial estimates of the parameters, as was the case with some of the model functions examined in Secs. 8.2 and 8.3. An adequate procedure might be to start with an estimate of γ, denoted γ_0, equal to .025 or .05 (say), and to solve the following system of linear least-squares equations resulting from model function (5.1):

$$n\alpha_0 - \beta_0 \sum \gamma_0^X = \sum Y$$
$$\alpha_0 \sum \gamma_0^X - \beta_0 \sum \gamma_0^{2X} = \sum \gamma_0^X Y \tag{8.69}$$

For the set of estimates α_0, β_0, and γ_0 obtained, the residual sum of squares RSS is given by (8.67) for the additive error assumption and by (8.68) for the multiplicative error assumption. Then γ_0 is incremented in steps of .025 or .05 (say) and the system of linear equations (8.69) solved and RSS reevaluated at each step. Unless the optimum value of γ is less than the initially used trial value of γ_0, the residual sum of squares RSS will decrease monotonically as γ_0 is increased until γ_0 is near its optimum value, after which RSS will increase. The set of estimates α_0, β_0, and γ_0 corresponding to the smallest RSS observed should provide good initial estimates for the Gauss–Newton procedure.

8.5 Summary

The reader will now appreciate that there are, in general, many methods
for obtaining initial estimates of the parameters of nonlinear
regression models for use as starting values with the Gauss–Newton
algorithm, when the latter is used to obtain the LS estimates. Clearly,
the more closely a model function/data set combination behaves like
a linear model, the less accurate the initial estimates need be, as
convergence is likely to occur without difficulty anyway. The more
nonlinear the behavior of the model function/data set combination, the
more stringent will be the requirements for good initial parameter
estimates. The procedures illustrated here are but a sample of the
total possibilities that may exist. The reader should be flexible and
be prepared to use more than one possibility. In some cases a mixture
of methods may be practicable (e.g., in Sec. 8.3.5 an estimate of γ
can be readily obtained from either an estimate of X_F or of $X_{1/2}$).
Modelers may sometimes have to try several alternatives in their quest
for satisfactory initial parameter estimates.

For simulation studies, it is usually sufficient to use as initial
estimates the known "true" parameter values that were employed to
generate the data. This often enables convergence to be obtained for
each of the 1000 (say) data sets generated, except for the most
nonlinear-behaving models such as the Bleasdale–Nelder and
Farazdaghi–Harris models, for example. Fortunately, however, rather
good readily programmable weighted least-squares methods are available
for these models (see Sec. 8.2), which may be used to obtain good initial
estimates for subsequent use with the Gauss–Newton algorithm.

In the absence of a procedure which provides good initial
estimates that virtually guarantees convergence, the user may have no
recourse other than to carry out the simulation study with a smaller
variance than the variance estimate $\hat{\sigma}^2$ obtained from the original data
set. The curvature measures of nonlinearity of Bates and Watts (1980)
are proportional to the square root of the variance. In addition, the
bias in the LS estimates using the calculation of Box (1971) is

proportional to the variance [see expression (2.3)]. Hence, as the variance is decreased, the model function/data set combination may be expected to behave more and more close to linear, and therefore convergence of the Gauss–Newton algorithm must eventually be achieved even with relatively poor initial estimates provided that the variance is chosen to be sufficiently small. Naturally, the user will prefer to carry out the simulation study employing the original variance, as only then can the properties and behavior of the LS estimators be fully assessed. Nevertheless, the set of estimates obtained from a simulation study employing a reduced variance still can tell the user which parameters are the ones giving rise to badly behaved LS estimators, whether the skewness is positive or negative, and may provide some indications as to a suitable reparameterization.

Exercises

8.1   Examine expression (8.10), which is the weighted least–squares criterion for the error assumption Var $(\log Y_t) = \sigma^2$. Determine the form that $E(Y_t)$ must have in order that (8.10) gives rise to a system of linear least–squares equations. List the models of Chapters 3–6 which are of this form. List the models of Chapters 3–6 which are of this form if one parameter is held constant.

8.2   Derive expressions (8.13), (8.20), (8.28), and (8.40).

8.3   Suppose that one were interested in obtaining initial estimates for model function (4.9) of the logistic model. Derive formulas relating the parameters of this model function to those of model function (4.8), for which a weighted least–squares procedure for obtaining good initial estimates is described in Sec. 8.3.2.

8.4   Using the second derivative with respect to X of model function (4.4) of the MMF model, show that the coordinates of the point of inflection of that model are given by expressions (8.51) and (8.52).

8.5   Using the second derivative with respect to X of model function

(4.5) of the Weibull-type model, show that the abscissa of the point of inflection of that model is given by expression (8.60).

8.6 Consider model function (6.2). Develop a method for obtaining good initial estimates of the parameters of this model function.

8.7 Consider model function (6.3), (6.5), or (6.6). Develop a method for obtaining good initial estimates of the parameters of these model functions which leads to a system of three linear equations.

8.8 Consider model function (6.7). Recognizing the similarity between this model function and model function (5.2), develop a method for obtaining good initial estimates of the parameters of this model function.

8.9 Consider model function (6.11). What is the relationship of this model function to that of Exercise 8.6? Develop a method for obtaining good initial estimates of the parameters of this model function. Now consider model function (6.10) and do the same for that.

9

Summary: Toward a Unified Approach to Nonlinear Regression Modeling

9.1 Introduction

Throughout this book there has been an underlying unifying theme based on the desirability of finding nonlinear regression models whose behavior in estimation approaches that of linear models. The use of linear models as a yardstick for measuring the performance of nonlinear models seems a natural one since in large samples, the behavior of nonlinear models in estimation approaches that of a linear model. That is, with the usual assumption about the stochastic term the LS estimators of the parameters of the model approach the condition of being unbiased, normally distributed, minimum variance estimators. The extent to which these properties hold in small samples thus seems a natural measure of the extent of nonlinearity in a nonlinear model. It is clear from the numerous examples in Chapters 3–6 that there is no way of defining a priori what constitutes a "large" sample. For some models studied in those chapters, the asymptotic properties were found to be closely approximated for sample sizes which are small with respect to the experimental effort required to collect the observations. Such models demonstrate what we have termed "close-to-linear" behavior, whereas other models only approach linear behavior for sample sizes which are well beyond the resources of the experimenter. In addition to increasing the sample size, a second way in which the asymptotic behavior may be approached is by reducing the residual variance. This follows from the fact that the curvature measures of nonlinearity of Bates and Watts are directly proportional to the square root of the error variance, and the bias in the LS estimates, as measured by Box's calculation, is directly proportional to the error variance. To use these measures in practice, of course, one substitutes the observed residual variance in place of the unknown true error variance. The smaller the investigator can make the residual variance, by careful and judicious experimental techniques, the closer will be the approach to linear behavior. Experimenters, however, may find it difficult to reduce the residual variance below some practical limit just as they will find that there is a maximum sample size in practice above which it is not feasible or economical to perform the experiment.

Readers are reminded that in the preface to this book it was pointed out that a user usually tries to fit a nonlinear model for one of three purposes: (1) to obtain a "good fit" to the data merely for purposes of representation, (2) to predict response values Y at given fixed values of the regressor variable X and to establish a confidence interval for those predicted values, and (3) to compare results obtained under different experimental conditions through use of, and interpretation of, the parameters. For the first of these purposes, which is outside the scope of this book, the user may be content to draw a freehand curve, or may seek more formal methodology such as the use of "spline functions" (e.g., see Wold, 1974). The second and third purposes are the realm of the present book, and Sec. 9.2 is devoted to a discussion of prediction and its relationship to the intrinsic nonlinearity of a model, whereas Sec. 9.3 is devoted to a discussion of parameter interpretation and its relationship to parameter-effects nonlinearity. Section 9.4 brings to the reader's attention a couple of commonly held fallacies in nonlinear regression modeling. Section 9.5 summarizes the recommended steps that the modeler should use to examine a model/data set combination for nonlinear behavior.

9.2 Consequences of Intrinsic Nonlinearity

In Chapters 1 and 2 the concept of "intrinsic" nonlinearity as used by Beale (1960) and Bates and Watts (1980) was introduced and discussed.

Readers will recall that this component of nonlinearity relates to the curvature of the solution locus (Sec. 2.4) and becomes defined when a model function/data set combination is specified. For most models that this author has investigated, the intrinsic nonlinearity has only rarely fallen outside acceptable limits, a result in agreement with that of Bates and Watts (1980), who found only four model/data set combinations with significant intrinsic curvatures out of 24 examples studied. The importance of intrinsic nonlinearity becomes manifest when the user wishes to predict values of the response variate Y and to determine confidence limits for those predicted values. For this objective, of the two components of nonlinearity for which Bates and Watts (1980) have provided measures, only the intrinsic component is relevant, as the following theoretical considerations and derivations will show.

Consider the following expression for the predicted value $\hat{Y}$ in a nonlinear regression model:

$$\hat{Y} = f(\underset{\sim}{X}, \hat{\underset{\sim}{\theta}}) \tag{9.1}$$

where the regressor variable vector $\underset{\sim}{X}$ is, in general, k-dimensional, although k = 1 for all but one example (see Sec. 6.3) treated in this book. In expression (9.1), $\hat{\underset{\sim}{\theta}}$ represents the vector of LS estimators of the parameters $\underset{\sim}{\theta} = (\theta_1, \theta_2, \ldots, \theta_p)^T$ and $\hat{Y}$ is the fitted or predicted value of the response Y corresponding to $\hat{\underset{\sim}{\theta}}$ at a fixed value of $\underset{\sim}{X}$. It is important to realize that the predicted value $\hat{Y}$ for a given $\underset{\sim}{X}$ will remain unaltered as a result of reparameterization of (9.1); any other model function of that model, obtained by reparameterizations of the form

$$\underset{\sim}{\phi} = g(\underset{\sim}{\theta}) \tag{9.2}$$

where the "new" parameter vector $\underset{\sim}{\phi}$ is a function only of the "old" parameters $\underset{\sim}{\theta}$ and not of $\underset{\sim}{X}$, results in exactly the same $\hat{Y}$ value as before (except for certain pathological cases of no practical interest). This is a consequence of the fact that the LS estimates of parameters expressed as a set of single-valued functions g of some other set of

parameters, are nothing more than g evaluated at the LS estimates of
the original parameters, that is,

$$\hat{\phi} = g(\hat{\theta}) \tag{9.3}$$

so that expression (9.1) may be written as

$$\hat{Y} = f[X, \, g^{-1}(\hat{\phi})] \tag{9.4}$$

Since the predicted value $\hat{Y}$ is independent of the parameterization,
it follows that any bias in $\hat{Y}$ cannot be the result of the
parameter-effects nonlinearity, which may, in principle if not always
in practice, be drastically reduced by a suitable reparameterization.
Therefore, the bias in $\hat{Y}$ must be related only to the intrinsic component
of nonlinearity. Box (1971, Sec. 6), making use of the first- and
second-order terms of the Taylor series expansion, derived the
relationship for predicting bias in $\hat{Y}$ from the calculated bias in the
parameter estimates as

$$\text{Bias } (\hat{Y}) = F^T \text{ Bias } (\hat{\theta}) + \frac{1}{2} \text{ tr } [H \text{ Cov } (\hat{\theta})] \tag{9.5}$$

where F is the p × 1 vector of the first derivatives of $f(X, \theta)$ and
H is the p × p matrix of second derivatives with respect to θ, these
derivatives being evaluated at $\hat{\theta}$. In addition, Bias $(\hat{\theta})$ is the p × 1
vector of biases from (2.3), and Cov $(\hat{\theta})$ is the p × p covariance matrix
of $\hat{\theta}$, obtained from the inverse of the information matrix (2.2). The
reader will note the close similarity of form between the right-hand
sides of (9.5) and (2.5), the difference being that (9.5) involves
derivatives of the model function $f(X, \theta)$, whereas (2.5) involves
derivatives of a reparameterization function $g(\theta)$.

Box (1971) also derived the formula for the variance of the
predicted value $\hat{Y}$, albeit to a lower order of accuracy than the bias
in $\hat{Y}$, as

$$\text{Var } (\hat{Y}) = F^T \text{ Cov } (\hat{\theta}) \cdot F \tag{9.6}$$

To illustrate the use of (9.5) and (9.6), consider the simple data set of Table 1.1 in combination with model function (1.14),

$$E(Y) = X^\theta$$

From Sec. 2.5, the following values were obtained:

$$\hat{\theta} = 2.0537$$
$$\text{Bias } (\hat{\theta}) = -.01327$$
$$\text{Var } (\hat{\theta}) = .02480$$

Differentiating the right-hand side of (1.14) with respect to θ, one obtains

$$F^T = F = (\log X)X^\theta$$

and

$$H = (\log X)^2 X^\theta$$

To predict the response Y corresponding to X = 2.5, for example, a point midway between the two X values of the original data set of Table 1.1,

$$\hat{Y} = (2.5)^{2.0537} = 6.5652$$
$$F^T = F = (\log 2.5)(2.5)^{2.0537} = 6.0156$$
$$H = (\log 2.5)^2(2.5)^{2.0537} = 5.5120$$

From (9.5), the bias in the prediction is

$$\text{Bias } (\hat{Y}) = 6.0156(-.01327) + \frac{1}{2}(5.5120)(.02480) = -.01148$$

and from (9.6), the variance of the prediction is

$$\text{Var } (\hat{Y}) = (6.0156)^2(.02480) = .8975$$

It is left to the reader as an exercise (see Exercise 9.1) to demonstrate that using the reparameterized model (1.15),

$$E(Y) = X^{\log \phi}$$

results in identical values of the bias and variance of the prediction $\hat{Y}$ as obtained above.

The bias in $\hat{Y}$ in the example above is relatively small, being only
$-.175\%$ of $\hat{Y}$. This is consistent with the fact that the intrinsic
nonlinearity was acceptably small for this model/data set combination
(see Sec. 2.4). The greater the curvature of the solution locus, the
greater will be the expected bias in the prediction of Y-values using
(9.5). This fact demonstrates that if the modeler's chief objective
is to predict values of the response variate Y, it will be necessary
to use a model in which the intrinsic nonlinearity falls within
acceptable limits. Most of the models studied in Chapters 3-6 have
acceptable intrinsic nonlinearities, although there were varying
degrees of acceptability for different models. For example, in Chapter
3, the Holliday model (3.2) has a very low intrinsic nonlinearity for
all five data sets studied, whereas the Bleasdale-Nelder model (3.1)
has a significant intrinsic nonlinearity for one of the data sets, and
is much closer to the critical values than the Holliday model (3.2)
for the other data sets. This higher intrinsic nonlinearity translates
itself into greater bias in the predictor $\hat{Y}$ than the alternative model,
as can be seen from the results (Table 9.1) of a calculation using
expression (9.5) for the TAS onion data set. At each of three densities,
the bias in the prediction is less with the Holliday model than with
the Bleasdale-Nelder model.

The sigmoidal growth models studied in Chapter 4 exhibit a wide
range of intrinsic curvatures, the extremes ranging from the logistic

Table 9.1 Bias in $\hat{Y}$ from (9.5) at Densities X = 10,105,200 (the Two
Extreme Values and Their Mean) for the TAS Onion Data of Chapter 3

Density	$\hat{Y}$	Bias ($\hat{Y}$) Bleasdale-Nelder (3.1)	Holliday (3.2)
10	5.812	.00163	.00035
105	4.313	.00075	.00006
200	3.738	-.00100	.00015

model (4.2), which has acceptably low intrinsic nonlinearities for each of the four data sets, to the Richards model (4.3), which has unacceptably high intrinsic nonlinearities for each of the four data sets.

In Chapter 5 the asymptotic regression model was studied, and it was found (see Table 5.2) that the intrinsic curvature exceeds the critical value for acceptability in four of the seven data sets studied. The significant curvature can be largely attributed to the extremely small sample size of most of the data sets (see Exercise 5.1). Altering the position of the X-values has only a minor effect on the intrinsic curvature (Table 5.3). Although the intrinsic curvature is never greatly in excess of the critical value, the user may expect to get some significant bias in predicted Y values using this model for small sample sizes.

Chapter 6 studied a variety of miscellaneous models. Acceptable intrinsic curvatures were obtained for all models in this chapter, except for the bent-hyperbola regression models of Sec. 6.6 [model functions (6.12) and (6.14)], where significant nonlinearities were obtained for the third of the three data sets and where the intrinsic curvature was close to significance for the first data set. The relatively high intrinsic nonlinearities of the bent-hyperbola models are not attributable to small sample sizes, as these ranged between 25 and 29. Thus it must be concluded that the tendency toward a significantly curved solution locus is an inherent feature of these models.

9.3 Consequences of Parameter-Effects Nonlinearity

In Chapter 2 the concept of "parameter-effects" nonlinearity as used by Bates and Watts (1980) was discussed (Sec. 2.4). In the 24 model/data set combinations that they studied, this component of nonlinearity always exceeded the intrinsic component of nonlinearity, often greatly. The examples studied in this book support that finding, as the parameter-effects curvature was greater than the intrinsic curvature in every example. However, the parameter-effects curvatures can be reduced, sometimes dramatically, by suitable reparameterization so as to bring them within acceptable limits, as some of the examples in Chapters 4 and 6 have shown. As pointed out by Bates and Watts (1980), and as is also clear from the material in Sec. 9.2, there will be little purpose in seeking reparameterizations if the intrinsic curvature is greatly outside acceptable limits, as any new model function will still exhibit significant nonlinear behavior in estimation no matter how good the reparameterization. Fortunately, however, a large majority of model/data set combinations that have been studied to date *do* have acceptably low intrinsic curvatures, which gives the search for more suitable parameterizations a sound practical basis.

Given that the intrinsic curvature is acceptably low, and if the modeler is successful in searching for a reparameterization which has an acceptable parameter-effects curvature, the modeler should then have a model function whose estimation behavior closely approximates that of a linear model. The benefits to the modeler of such a model function will be as follows:

1. The LS estimates will be easily obtained using the Gauss–Newton algorithm, a program for which appears in the Appendix to this book. For a linear model, convergence to the LS estimates (which are also the ML estimates under the assumption of iidN error) occurs in a single iteration from any set of initial parameter estimates (see Appendix 2.A). The closer a nonlinear model approaches a linear model in behavior, the fewer is the number of iterations needed for convergence to be obtained, even from crude initial estimates remote from the LS estimates. This may be contrasted with badly behaved nonlinear models which may not converge to the LS estimates at all, unless very good initial estimates (i.e., ones close to the LS estimates) can be obtained.

2. Because the LS estimators will be almost unbiased and almost normally distributed, and because the actual variances of the estimates will be close to those given by the asymptotic covariance matrix, this will enable crude preliminary comparisons such as t-tests of parameter estimates between data sets to be carried out. It is emphasized that these comparisons should be

viewed as preliminary ones, because the formal procedure for conducting such tests (see Chapter 7), based on the "extra sum of squares" principle, is affected *only* by the intrinsic nonlinearity.

The closer a nonlinear model approaches linear behavior, the more valid will be various statistical tests or procedures whose use is derived from an analogy with linear models and which assume normality. For these tests or procedures, the user can feel confident that the supposed significance levels are being closely approximated when the model is close to linear. For example, consider expression (2.9) for obtaining confidence regions for the parameters. The more linear the behavior of the model, the closer the confidence regions will be to ellipsoids (ellipses in two dimensions and hyperellipsoids in more than three dimensions). Furthermore, the actual but unknown significance level will agree more closely with the nominal significance level α as the intrinsic nonlinearity becomes smaller and smaller, since the bias in the residual sum of squares depends on the extent of intrinsic nonlinearity.

9.4 Some Fallacies in Nonlinear Regression Modeling

This section examines two commonly held fallacies in nonlinear regression modelling. The first involves the effect of correlation between parameters on the nonlinear behavior of the LS estimators, and the second involves the relationship between the manner in which the parameters appear in a model and how they behave in estimation. Each is considered in a separate subsection.

9.4.1 Parameter correlation and nonlinear behavior

It is fallacious that nonlinear behavior in estimation of nonlinear regression models (i.e., bias, excess variance, and non-normal distribution of the LS estimators of the parameters) may be attributable to, or diagnosed by, high correlation between the LS estimators of the parameters. That this is false can be seen by an examination of the most simple linear regression model, the "straight-line" regression model

$$E(Y) = \alpha + \beta X \qquad (9.7)$$

Since (9.7) is a linear model, the LS estimates of the parameters α and β can be explicitly obtained by minimizing the sum of squares

$$S = \sum_{t=1}^{n} (Y_t - \alpha - \beta X_t)^2 \qquad (9.8)$$

to yield

$$\hat{\alpha} = \frac{\sum Y_t}{n}$$

$$\hat{\beta} = \frac{\sum X_t Y_t - (\sum X_t)(\sum Y_t)/n}{\sum X_t^2 - (\sum X_t)^2/n}$$

Denoting the vector of parameters by $\underset{\sim}{\theta}$ $(= \alpha, \beta)$, the information matrix (2.2) is

$$\frac{1}{\sigma^2} \begin{pmatrix} n & \sum X_t \\ \sum X_t & \sum X_t^2 \end{pmatrix}$$

so that the covariance matrix of $\hat{\alpha}$ and $\hat{\beta}$, which is the inverse of the information matrix, is

$$\text{Cov}\,(\underset{\sim}{\hat{\theta}}) = \frac{\sigma^2}{n \sum X_t^2 - (\sum X_t)^2} \begin{pmatrix} \sum X_t^2 & -\sum X_t \\ -\sum X_t & n \end{pmatrix}$$

from which the sample correlation coefficient r between $\hat{\alpha}$ and $\hat{\beta}$ is simply

$$r = \frac{-\sum X_t}{(n \sum X_t^2)^{1/2}} = \frac{-\bar{X}}{(\sum X_t^2/n)^{1/2}}$$

It may easily be verified that the further the sample mean $\bar{X}$ of the X's is from the origin relative to the variance of the X's, the higher is the absolute value of the correlation coefficient. Or, stated equivalently, the smaller the coefficient of variation of the X's, the higher is $|r|$. For example, if n = 4 and X_t = (1, 2, 3, 4), r = −.913, but if X_t = (101, 102, 103, 104), r = −.99994, and if X_t = (1001, 1002, 1003, 1004), r = −.9999994. Thus it is clear that the correlation between $\hat{\alpha}$ and $\hat{\beta}$ can be made as large as possible merely by a location change of the X−values. Since (9.7) is a linear model, then under the assumption of an iidN stochastic term, $\hat{\alpha}$ and $\hat{\beta}$ must be jointly normally distributed, unbiased, minimum variance estimators. High correlation between the LS estimators in nonlinear regression models thus cannot, in general, be indicative or diagnostic of drastic nonlinear behavior, as equivalent high correlations may be obtained in linear models where the LS estimators are known to have the most desirable properties that estimators may have. Perhaps the reason for the origin of the fallacy lies in the fact that in badly behaved nonlinear models, the correlations between parameter estimators are often observed to be high. However, there are also examples of badly behaved nonlinear models in which the correlations are not particularly high. We now present a few illustrations of sample correlations between parameter estimators for some of the examples considered in this book which help illustrate the points noted above.

Consider the logistic model in model function (4.2),

$$Y = \frac{\alpha}{1 + \exp(\beta - \gamma X)}$$

which was found to approach linear behavior in estimation sufficiently closely for practical use. The correlation coefficients between $\hat{\alpha}$, $\hat{\beta}$, and $\hat{\gamma}$ are as follows for the "onion bulbs plus tops" data set:

	α	β	γ
α	1.0		
β	−.405	1.0	
γ	−.544	.964	1.0

The relatively high correlation of .964 between $\hat{\beta}$ and $\hat{\gamma}$ is not reflected in measurable nonlinear behavior of the estimators of β and γ.

Consider now the Weibull-type model in model function (4.5):

$$Y = \alpha - \beta \exp(-\gamma X^\delta)$$

The matrix of correlation coefficients between parameter estimators for the "pasture regrowth" data set is as follows:

	α	β	γ	δ
α	1.0			
β	.925	1.0		
γ	.710	.862	1.0	
δ	-.766	-.891	-.996	1.0

The correlation coefficient between $\hat{\gamma}$ and $\hat{\delta}$ is thus seen to be rather high. Reparameterization of (4.5) to give model function (4.15),

$$Y = \alpha - \beta \exp[-\exp(-\gamma)X^\delta]$$

results in much closer to linear behavior than that of the original model function considered (see Sec. 4.4.5). The matrix of correlation coefficients between the LS estimators of the parameters of (4.15) for "pasture regrowth" is as follows:

	α	β	γ	δ
α	1.0			
β	.925	1.0		
γ	-.710	-.862	1.0	
δ	-.766	-.891	.996	1.0

The only difference between parameterizations is that the signs of all correlations involving parameter γ have changed, but the magnitudes have not. This result has its parallel in many other examples using

the models and data of Chapters 3–6, where reparameterization results
at most in a change of sign of the correlation coefficient but not in
the magnitude. A further example of this is given in Exercise 9.4.

Consider now the correlation matrices for the parameter
estimators for the Richards model in model function (4.3) and the
generalized von Bertalanffy model in model function (4.13) for the
pasture regrowth data set:

Model Function (4.3)

	α	β	γ	δ
α	1.0			
β	−.820	1.0		
γ	−.880	.990	1.0	
δ	−.804	.996	.978	1.0

Model Function (4.13)

	α	β	γ	δ
α	1.0			
β	.990	1.0		
γ	−.977	−.985	1.0	
δ	.982	.956	−.965	1.0

The correlation coefficients are seen to be relatively high for both
model functions, both of which exhibit highly nonlinear behavior
(Section 4.4.3). However, no coefficient is higher than the highest
coefficient .996 observed for the reparameterized Weibull–type model
in model function (4.15), which behaves close to linear.

A final illustration will be provided by the badly behaved
Bleasdale–Nelder model (3.1) and the well–behaved Holliday model
(3.2). Correlation matrices are presented below for the parameter

estimators of each of these models for the Mt. Gambier data set on onions:

Bleasdale-Nelder Model (3.1)

	α	β	θ
α	1.0		
β	.9967	1.0	
θ	-.9998	-.9981	1.0

Holliday Model (3.2)

	α	β	γ
α	1.0		
β	-.9612	1.0	
γ	.8806	-.9647	1.0

Although the correlations between parameter estimators are higher for the Bleasdale-Nelder model than they are for the Holliday model, the user would have had no real indication of the poor distributional properties of the LS estimators of the parameters of the former model compared with the good distributional properties of the LS estimators of the parameters of the latter model. Hence the magnitude of correlation coefficients between parameter estimators is of no use as an indicator of the extent of nonlinear behavior of LS estimators.

9.4.2 Linear-appearing and nonlinear-appearing parameters

In Sec. 1.1 the distinction was made between linear and nonlinear regression models by defining a linear model as one that is linear in its parameters, meaning *all* its parameters. A nonlinear model is one in which at least one of the parameters appears nonlinearly. However, the question of what exactly is meant by an individual parameter appearing linearly is somewhat ambiguous. It is easier to discuss this by reference to an example; therefore, consider the asymptotic regression model that was studied in Chapter 5, in model function (5.1):

$$Y = \alpha - \beta\gamma^X$$

A "linear" parameter, or "linear-appearing" parameter, may be defined as one for which the derivative of the model function with respect to the parameter is independent of that parameter, *or* it may be defined as one for which the derivative is independent of *any* parameter. By either definition, α in (5.1) is a linear-appearing parameter, whereas β would be considered to be linear by the first definition, but not by the second. Some authors such as Lawton and Sylvestre (1971) find value in the concept of linear parameters in the development of algorithms for finding the LS estimates of the parameters. For example, if γ is held constant in (5.1), the model becomes linear in α and β irrespective of which of the foregoing definitions is used. We have made extensive use of this same principle in Chapter 8 in the search for good initial parameter estimates.

However, it is important to realize that whether or not a parameter appears linearly in a nonlinear regression model, this bears no relationship to its behavior in estimation. In model function (5.1), parameter α, which is linear appearing by either definition, and parameter β, which is linear appearing by one of the definitions, were more nonlinear in their estimation behavior than parameter γ, which is a nonlinear-appearing parameter by either definition. Box (1971, p. 176) noted that, if there are k responses rather than a single response as has been the case throughout this book, and if only one of these responses is nonlinear in the parameters, all the remaining k - 1 responses being linear in the parameters, then, in general, *all* the parameter estimates are biased. The degree of bias may be greater for the linear-appearing parameters than for the nonlinear-appearing parameters. Numerous examples can be offered to demonstrate that there is no necessary connection between linear appearance of a parameter in a model and the degree of nonlinear behavior of the estimators of that parameter.

9.5   Recommendations to the Modeler on the Procedure for Examining Nonlinear Behavior

The basic aim of this book has been to instruct the reader and potential user of nonlinear regression models about how to examine a model/data set combination for nonlinear behavior. The theme has been to search for models or model functions whose least-squares parameter estimators approach closely the condition of being unbiased, normally distributed, and having minimum variance, the properties expected for linear models. However, in order to search for models exhibiting close-to-linear behavior it is essential to have a systematic procedure for measuring the degree of departure of a model/data set combination from linear behavior. Chapter 2 was devoted to this question.

Clearly, the first step in an assessment of nonlinear behavior is to obtain the LS estimates. The speed of convergence of the Gauss–Newton algorithm, and the question of how sophisticated the procedure need be for obtaining initial parameter estimates, may themselves shed light on how closely the model approaches linear behavior. As discussed in Chapters 2 and 8, fast convergence from initial estimates far from the LS estimates will almost always indicate a model that is close to behaving like a linear model. One the other hand, slow convergence, especially from initial estimates reasonably close to the LS estimates, will almost certainly indicate that the LS estimators are far from linear in their estimation properties; that is, they will be significantly biased and non-normally distributed.

Having obtained the LS estimates, the following procedures should be carried out, as has been described in detail in Chapter 2, these being the most efficacious techniques for identifying and revealing nonlinear behavior:

1. Curvature measures of nonlinearity of Bates and Watts (Sec. 2.4)
2. Bias calculation of M. J. Box (Sec. 2.5)
3. Simulation studies (Secs. 2.6 and 2.9)

All of the procedures above have the common feature of employing the X-values (i.e., the values of the "regressor" or "predictor" variable) of the original data set, the LS estimates $\hat{\underset{\sim}{\theta}}$ in place of the parameter vector $\underset{\sim}{\theta}$, and the estimated residual variance $\hat{\sigma}^2$ in place of the unknown error variance σ^2.

Of the two approaches to assessing nonlinear behavior of nonlinear regression models noted in Sec. 1.4, the use of the curvature measures of Bates and Watts and the bias calculation of Box constitutes the "analytical" approach, one in which mathematical formulas are brought to bear upon a model/data set combination after the LS estimates of the parameters have been determined. The use of simulation studies constitutes a second and more "empirical" approach to assessing nonlinear behavior through an examination of the sampling properties of the LS estimator. In an ideal world, agreement would always be achieved between the two approaches. Indeed, this has often been the case in many of the examples that have appeared in the text and exercises in Chapters 3–6. For example, consider the Holliday model (3.2) used for the yield-density studies of Chapter 3. Both the intrinsic and parameter-effects components of the curvature measures of nonlinearity indicate that this model exhibits close-to-linear behavior for all data sets studied. In addition, Box's bias calculation shows that there is very little bias in the LS estimators, and finally, simulation studies confirm that the estimators are close to being unbiased and normally distributed with variances close to the minimum attainable variance.

However, agreement between the two approaches was not achieved for all models in this book. The most notable disagreement occurred for the model of Sec. 6.5 relating the resistance of a thermistor to temperature, where the curvature measure of parameter-effects nonlinearity was highly significant but a simulation study indicated close-to-linear behavior. A further contradictory result was obtained with the bent-hyperbola regression models of Sec. 6.6. Simulation studies show that the behavior of a reparameterization involving one of the five parameters is better than that of the original formulation [model function (6.13) versus (6.12)], but the curvature measures of parameter-effects nonlinearity are greater for the new model function.

The reasons for the failure of the two approaches to agree on some occasions is at present not understood. To resolve this contradiction one would have to examine geometrically the five-dimensional parameter space.

When genuinely significant parameter-effects nonlinearity is present, provided that the intrinsic nonlinearity is acceptably low, reparameterizations may then be sought to produce a model function having parameters whose LS estimators behave much more linearly than those of the parameters in the original parameterization. In Sec. 2.5, expression (2.5) was presented, which enables one to calculate the bias in the LS estimator of a parameter of a new parameterization from the biases in the LS estimators of the parameters of the old parameterization. This formula is useful for making preliminary checks as to whether some suggested reparameterization may be of value. To check the efficacy of any suggested reparameterization, one should carry out a simulation study. Such simulation studies need not be costly nor wasteful of computer time. If, as suggested in Sec. 2.6, an appropriate size for a simulation study involves generating 1000 sets of data, the parameter estimates from each of these sets can be stored on tape or disk. For model functions of the same model, each new parameter may be represented by a function of the old parameters, as

$$\phi = g(\underset{\sim}{\theta}) \tag{9.9}$$

that is, the new parameter ϕ is a function of one or more of the elements θ_j of $\underset{\sim}{\theta}$ ($j = 1, 2, \ldots, p$). A set of 1000 values of $\hat{\phi}$ can be obtained merely by substituting the saved parameter estimates of $\underset{\sim}{\theta}$ into (9.9); that is, one uses

$$\hat{\phi} = g(\hat{\underset{\sim}{\theta}})$$

By this means, a wide variety of alternative reparameterizations may be studied inexpensively using only the original set of estimates $\hat{\underset{\sim}{\theta}}$.

In addition to the three major techniques mentioned above (curvature measures of Bates and Watts; Box's bias measure; simulation studies), Chapter 2 dealt with two other techniques. These involve constructing confidence regions for the parameters (Sec. 2.7) and examining t-values for each parameter estimate (Sec. 2.8). The former was seen to be of limited general value because, in addition to lacking any formal test of significance for adjudging whether the contours depart from "ellipticity," the visual assessment cannot, in our three-dimensional world, be carried out for models having more than three parameters, and is very expensive of computing time even for two or three parameters. The use of t-values may be of some assistance in evaluating nonlinear behavior, as was shown by an example of its use in Chapter 4 (Sec. 4.5 and Table 4.9). A low t-value, which is the ratio of the parameter estimate to its standard error, usually indicates that the parameter is badly determined, although a t-value may sometimes be low because the estimate is near zero. A high t-value, on the other hand, while tending to indicate that a parameter is well determined, does not necessarily mean that the corresponding LS estimator exhibits close-to-linear behavior, as several examples have indicated otherwise [e.g., see Sec. 6.6, parameter θ_1 in model functions (6.12) and (6.14); Chapter 5, parameter α in model function (5.1); Sec. 2.8, parameter θ in model function (1.14)]. However, in all but the last example cited, the parameter in question represents a "constant" term in the model function in which it appears. As discussed in Sec. 6.6, the magnitude of the t-values may be made arbitrarily large or small by location changes of the response variable Y. Because of this, it is clear that a t-value cannot be of use for judging the estimation behavior of a constant term in a model function.

A similar problem occurs with the use of "percentage bias" (Sec. 2.5). For most models studied in this book, a percentage bias in a parameter estimate of less than 1% has usually correlated closely with close-to-linear behavior of the estimator. However, for a parameter representing a constant term in a model function, the percentage bias, like the t-value discussed in the preceding paragraph, can be made arbitrarily large or small. Excluding the case of parameters representing constant terms in model functions, the percentage bias does appear to be of assistance in helping to identify which of the parameters are the main contributors to the nonlinear behavior of the model function. Clearly, a parameter whose estimates have a high percentage bias must be indicative of a poorly formulated model for estimation.

Because the study of the behavior of the LS estimates in nonlinear models is a relatively new field, it seems wise to take a conservative approach to the question of whether or not a positive recommendation should be made about the use of particular models for given types of data. Only with the passage of time and the continued use of the methodology espoused herein, or with new theoretical developments which are bound to result from an increasing number of statisticians interesting themselves in the question of the properties of LS estimators, will a less piecemeal picture emerge, and perhaps then more sweeping generalizations may be made about the efficacy of particular models. Until such time, one can only recommend the use in conjunction of several of the techniques described in this book. To assist the user in making direct and immediate use of the material, the Appendix to this book contains listings of computer programs and detailed examples of how to use those programs. This should enable the modeler to put into practice the procedures advocated herein. As with every new field of endeavor, the time and effort spent at this early stage is bound to be rewarding and sometimes even exciting.

Exercises

9.1 Calculate the bias and variance of $\hat{Y}$ using expressions (9.5) and (9.6) for model function (1.15) in combination with the data of Table 1.1, making use of the numerical values of Bias ($\hat{\phi}$) and Var ($\hat{\phi}$) in Sec. 2.5, and show that the bias and variance of $\hat{Y}$ corresponding to X = 2.5 are identical to those calculated for model function (1.14) in Sec. 9.2.

9.2 Consider expression (9.5) and prove that for a linear model, Bias $(\hat{Y}) = 0$.

9.3 Use expression (9.5) to calculate the bias in the prediction $\hat{Y}$ corresponding to the densities X = 21, 80, and 155, respectively, for the Mt. Gambier onion data set of Appendix 3.A, in combination with the Bleasdale-Nelder (3.1), Holliday (3.2) and Farazdaghi-Harris (3.3) yield-density models in their logarithmically transformed forms. Do the relative magnitudes of the biases from the three models accord with what you would expect from the curvature measures of intrinsic nonlinearity reported in Table 3.1? To perform the calculation, you will need the covariance matrix of the parameter estimates, and their biases. Use the programs in the Appendix to obtain these.

9.4 The correlation matrix for the parameter estimators in the Morgan-Mercer-Flodin model function (4.4) originally studied in Chapter 4 was identical to that of the reparameterized model function (4.14) for the "pasture regrowth" data set of Appendix 4.A. The correlation matrix common to both model functions is as follows:

Model Functions (4.4) and (4.14)

	α	β	γ	δ
α	1.0			
β	-.539	1.0		
γ	-.866	.763	1.0	
δ	-.907	.726	.995	1.0

Was there a difference in the estimation behavior of the two model functions, and if so, what does this tell us about the value of parameter correlations in helping to make deductions about the estimation behavior of model functions?

References

Bacon, D. W. and Watts, D. G. (1971). Estimating the Transition Between Two Intersecting Straight Lines, *Biometrika 58*, 525-534.

Bard, Y. (1974). *Nonlinear Parameter Estimation*, Academic, New York.

Bates, D. M. and Watts, D. G. (1980). Relative Curvature Measures of Nonlinearity, *J. R. Statist. Soc., Ser. B 42*, 1-25.

Beale, E. M. L. (1960). Confidence Regions in Nonlinear Estimation, *J. R. Statist. Soc., Ser. B 22*, 41-76.

Bleasdale, J. K. A. and Nelder, J. A. (1960). Plant Population and Crop Yield, *Nature 188*, 342.

Box, G. E. P. and Hunter, W. G. (1965). The Experimental Study of Physical Mechanisms, *Technometrics 7*, 23-42.

Box, G. E. P. and Lucas, H. L. (1959). Design of Experiments in Non-linear Situations, *Biometrika 46*, 77-90.

Box, M. J. (1971). Bias in Nonlinear Estimation, *J. R. Statist. Soc., Ser. B 33*, 171-201.

Causton, D. R. (1969). A Computer Program for Fitting the Richards Function, *Biometrics 25*, 401-409.

Chambers, J. (1973). Fitting Nonlinear Models: Numerical Techniques, *Biometrika 60*, 1-13.

Daniel, C. and Wood, F. S. (1971). *Fitting Equations to Data*, Wiley-Interscience, New York.

Davies, O. L. and Ku, J. Y. (1977). Re-examination of the Fitting of the Richards Growth Function, *Biometrics 33*, 546-547.

Draper, N. R. and Smith, H. (1981). *Applied Regression Analysis*, 2nd ed., Wiley, New York.

Dudzinski, M. L. and Mykytowycz, R. (1961). The Eye Lens as an Indicator of Age in the Wild Rabbit in Australia, *CSIRO Wildl. Res. 6*, 156-159.

Farazdaghi, H. and Harris, P. M. (1968). Plant Competition and Crop Yield, *Nature 217*, 289-290.

Gallant, A. R. (1976). Confidence Regions for the Parameters of a Nonlinear Regression Model, *Inst. Statist. Mimeogr. Ser. 1077*, North Carolina State University, Raleigh, N.C.

Gillis, P. R. and Ratkowsky, D. A. (1978). The Behaviour of Estimators of the Parameters of Various Yield-Density Relationships, *Biometrics 34*, 191-198.

Gregory, F. G. (1956). General Aspects of Leaf Growth, in F. L. Milthorpe (Ed.), *The Growth of Leaves*, Butterworth, London.

Griffiths, D. A. and Miller, A. J. (1973). Hyperbolic Regression - A Model Based on Two-Phase Piecewise Linear Regression with a Smooth Transition Between Regimes, *Commun. Statist. 2*, 561-569.

Gunst, R. F. and Mason, R. L. (1980). *Regression Analysis and Its Application: A Data-Oriented Approach*, Dekker, New York.

Guttman, I. and Meeter, D. A. (1965). On Beale's Measures of Non-linearity, *Technometrics 7*, 623-637.

Hartley, H. O. (1961). The Modified Gauss-Newton Method for the Fitting of Nonlinear Regression Functions by Least Squares, *Technometrics 3*, 269-280.

Heyes, J. K. and Brown, R. (1956). Growth and Cellular Differentiation, in F. L. Milthorpe (Ed.), *The Growth of Leaves*, Butterworth, London.

Hill, A. V. (1913). The Combinations of Haemoglobin with Oxygen and Carbon Monoxide, *Biochem. J. 7*, 471-480.

Holliday, R. (1960). Plant Population and Crop Yield, *Field Crop Abstr. 13*, 159-167, 247-254.

Jennrich, R. I. (1969). Asymptotic Properties of Non-linear Least Squares Estimators, *Ann. Math. Statist. 40*, 633-643.

Keeping, E. S. (1951). A Significance Test for Exponential Regression, *Ann. Math. Statist. 22*, 180-198.

Kendall, M. G. and Stuart, A. (1963). *The Advanced Theory of Statistics,* Vol. 1: *Distribution Theory*, 2nd ed., Charles Griffin, London.

Kendall, M. G. and Stuart, A. (1967). *The Advanced Theory of Statistics,* Vol. 2: *Inference and Relationship*, 2nd ed., Charles Griffin, London.

Kennedy, W. J. and Gentle, J. E. (1980). *Statistical Computing*, Dekker, New York.

Lawton, W. H. and Sylvestre, E. A. (1971). Elimination of Linear Parameters in Nonlinear Regression, *Technometrics 13*, 461-467.

Li, J. C. R. (1964). *Statistical Inference*, Vol. I, Edwards Brothers, Ann Arbor, Mich.

Malinvaud, E. (1970). *Statistical Methods of Econometrics*, 2nd ed., North-Holland, Amsterdam.

Marquardt, D. W. (1963). An Algorithm for Least Squares Estimation of Nonlinear Parameters, *J. Soc. Ind. Appl. Math. 2*, 431-441.

Marsh, H. (1980). Age Determination of the Dugong [*Dugong dugon* (Müller)] in Northern Australia and Its Biological Implications, in W. F. Perrin and A. C. Myrick (Eds.), *Age Determination in Toothed Cetaceans and Sirenians*, International Whaling Commission, Cambridge.

Mead, R. (1970). Plant Density and Crop Yield, *Appl. Statist. 19*, 64-81.

Mead, R. (1979). Competition Experiments, *Biometrics 35*, 41-54.

Meyer, R. R. and Roth, P. M. (1972). Modified Damped Least Squares: An Algorithm for Non-linear Estimation, *J. Inst. Math. Appl. 9*, 218-233.

Michaelis, L. and Menten, M. L. (1913). Die Kinetik der Invertinwirkung, *Biochem. Z. 49*, 333-369.

Milliken, G. A. and DeBruin, R. L. (1978). A Procedure to Test Hypotheses for Nonlinear Models, *Commun. Statist.-Theory Methods A7*, 65-79.

Morgan, P. H., Mercer, L. P., and Flodin, N. W. (1975). General Model for Nutritional Responses of Higher Organisms, *Proc. Nat. Acad. Sci. U.S.A. 72*, 4327-4331.

Nelder, J. A. (1963). Yield Density Relations and Jarvis' Lucerne Data, *J. Agric. Sci. 61*, 427-429.

Oliver, F. R. (1966). Aspects of Maximum Likelihood Estimation of the Logistic Growth Function, *J. Amer. Statist. Assoc. 61*, 697-705.

Phillips, B. F. and Campbell, N. A. (1968). A New Method of Fitting the von Bertalanffy Growth Curve Using Data on the Whelk *Dicathais*, *Growth 32*, 317-329.

Pimentel-Gomes, F. (1953). The Use of Mitscherlich's Regression Law in the Analysis of Experiments with Fertilizers, *Biometrics 9*, 498-516.

Radway Allen, K. (1966). A Method of Fitting Growth Curves of the von Bertalanffy Type to Observed Data, *J. Fish. Res. Board Can. 23*, 163-179.

Richards, F. J. (1959). A Flexible Growth Function for Empirical Use, *J. Exp. Biol. 10*, 290-300.

Sandland, R. L. and McGilchrist, C. A. (1979). Stochastic Growth Curve Analysis, *Biometrics 35*, 255-271.

Shaw, D. E. and Griffiths, D. A. (1979). Likelihood Ratio and Exact Confidence Regions in Nonlinear Regression, *Symposium on Regression Analysis*, Statistical Society of Australia, N.S.W. Branch, Aug. 30-31, 1979.

Shinozaki, K. and Kira, T. (1956). Intraspecific Competition Among Higher Plants, VII: Logistic Theory of the C-D Effect, *J. Inst. Polytech., Osaka City Univ., Ser. D 7*, 35-72.

Snedecor, G. W. and Cochran, W. G. (1967). *Statistical Methods*, 6th ed., Iowa State University Press, Ames, Iowa.

Snedecor, G. W. and Cochran, W. G. (1980). *Statistical Methods*, 7th ed., Iowa State University Press, Ames, Iowa.

Stevens, W. L. (1951). Asymptotic Regression, *Biometrics 7*, 247-267.

Von Bertalanffy, L. (1951). *Theoretische Biologie*, Band II, 2nd ed., A. Frank, Bern.

Watts, D. G. and Bacon, D. W. (1974). Using a Hyperbola as a Transition Model to Fit Two-Regime Straight-Line Data, *Technometrics 16*, 269-373.

Weibull, W. (1951). A Statistical Distribution Function of Wide Applicability, *J. Appl. Mech. 18*, 293-296.

Willey, R. W. and Heath, S. B. (1969). The Quantitative Relationships Between Plant Population and Crop Yield, *Adv. Agron. 21*, 281-321.

Williams, E. J. (1959). *Regression Analysis*, Wiley, New York.

Wold, S. (1974). Spline Functions in Data Analysis, *Technometrics 16*, 1-11.

Yang, R. C., Kozak, A., and Smith, J. H. G. (1978). The Potential of Weibull-Type Functions as Flexible Growth Curves, *Can. J. For. Res. 8*, 424-431.

Appendix

Summary

1. Introduction

2. Subroutines for general use. Four subroutines are listed. Full
   instructions on their use appear as a set of comments immediately
   after the SUBROUTINE statement. A short description of each of
   the subroutines follows:

   2-1. Subroutine SOLVE estimates the parameters for a nonlinear
        system by the Gauss–Newton method (Kennedy and Gentle,
        1980).

   2-2. Subroutine BATES calculates the Bates and Watts measures
        of maximum intrinsic and parameter-effects curvature
        (Bates and Watts, 1980) and Box's measure of bias in the
        parameter estimators (Box, 1971).

   2-3. Subroutine GNRATE returns a vector of simulated dependent
        variable values with independent and identically
        distributed normal error components.

   2-4. Subroutine SUMRY supplies various summary statistics for
        a vector of estimates for each parameter obtained through
        simulation.

3. Auxiliary subroutines. Eleven subroutines required by the general
   use subroutines detailed above are listed in alphabetical order.

4. Example programs. Two example programs are listed along with the output produced on a CDC CYBER 76 computer.

> 4-1. Program XAMPLE provides an example of the use of subroutines SOLVE, BATES, GNRATE, and SUMRY to obtain parameter estimates for a model/data set combination and to examine the properties of the parameter estimators.
>
> 4-2. Program COMPRE illustrates the use of subroutine SOLVE in the comparison of parameter estimates between different data sets by the extra sum of squares principle (Draper and Smith, 1981).

5. References

1. Introduction

This appendix contains listings of a collection of FORTRAN subroutines (referred to as the NONLIN subroutines) which enable the fitting of a nonlinear model to a data set and the subsequent investigation of the properties of the parameter estimators for that model/data set combination. Two example programs illustrating the use of the NONLIN subroutines are also listed in full, together with the output these programs produce when implemented on a CDC Cyber 76 computer. The NONLIN subroutines and the example programs have been written to comply with the USA Standard FORTRAN specifications (American National Standards Institute, 1966) and have been tested for portability using the program PFORT (Ryder and Hall, 1973). The three known nonportable features of the routines are as follows:

1. The program header card with file declarations is required by the CDC Cyber 76 FORTRAN Extended compiler and has been included as a comment card in the two example programs.

2. Subroutine RANSET and function RANF are CDC Cyber 76 FORTRAN Extended initialization and uniform pseudo-random number generator routines, respectively. They are to be replaced by similar routines available at the user's installation and are

flagged in the listings with comment cards with C*** in columns 1 to 4.

3. Parentheses in several FORMAT statements are nested to more than two levels. This is legal in USA Standard FORTRAN but may not be compatible with some FORTRAN compilers.

The NONLIN subroutines require access to a number of subroutines from the LINPACK package, full listings of which can be found in the LINPACK Users' Guide (Dongarra et al., 1979).

A magnetic tape containing the FORTRAN source code for the NONLIN subroutines and the example programs can be obtained from the following source:

Dr. D. A. Ratkowsky
CSIRO Division of Mathematics and Statistics
Tasmanian Regional Laboratory
Hobart   7000
Tasmania
Australia

Information regarding costs and tape formats can be obtained directly from the source above.

The NONLIN subroutines and examples have been checked for a number of model/data set combinations. However, the author accepts no responsibility for any errors that may result from the use of the FORTRAN source code supplied in the following listings or on magnetic tape.

2. Subroutines for General Use

```
      SUBROUTINE SOLVE(Y,PARAM,ALPHA,NITS,NPTS,NP,ND1,LUNOUT,JOB,EVAL,
     +             VARC,RSS,IFAIL,DERIVS,QRAUX,JPVT,DIFF,WK1)
C
C-----------------------------------------------------------------------
C
C     SUBROUTINE SOLVE ESTIMATES THE PARAMETERS FOR A NONLINEAR SYSTEM BY
C     THE GAUSS-NEWTON METHOD (KENNEDY AND GENTLE, 1980).
C
C     DESCRIPTION
C     THE USER-SUPPLIED SUBROUTINE EVAL IS CALLED NPTS TIMES TO CONSTRUCT
C     A MATRIX OF FIRST-ORDER PARTIAL DERIVATIVES BASED ON INITIAL
C     PARAMETER ESTIMATES PROVIDED BY THE USER.  A VECTOR OF DIFFERENCES
C     IS ALSO CONSTRUCTED BY SUBTRACTING PREDICTED VALUES OBTAINED FROM
C     THE USER-SUPPLIED SUBROUTINE EVAL FROM THE VALUES OF THE DEPENDENT
C     VARIABLE.  THE LINPACK ROUTINE SQRDC IS THEN USED TO DECOMPOSE THE
C     FIRST DERIVATIVE MATRIX (BUSINGER AND GOLUB, 1965) AND THE RESULTS OF
```

```
C       THIS QR DECOMPOSITION, ALONG WITH THE DIFFERENCE VECTOR, ARE USED
C       BY THE LINPACK ROUTINE SQRSL TO FIND CORRECTIONS TO THE PARAMETER
C       ESTIMATES.  THIS PROCEDURE IS REPEATED WITH THE NEW PARAMETER
C       ESTIMATES UNTIL CONVERGENCE OCCURS (PARAMETER CORRECTIONS ARE
C       SUFFICIENTLY SMALL) OR UNTIL A SPECIFIED MAXIMUM NUMBER OF ITERATIONS
C       IS EXCEEDED.  IF CONVERGENCE IS ACHIEVED, SUBROUTINE SOLVE THEN CALLS
C       SUBROUTINE DSPLAY TO PRINT RESULTS AS SPECIFIED BY THE ENTRY
C       ARGUMENT JOB.
C
C       ON ENTRY
C          (NOTE - ALL ARGUMENTS IN THE ENTRY LIST ARE RETURNED UNALTERED
C          UNLESS DETAILED IN THE EXIT LIST.)
C          Y      REAL(K), K.GE.NPTS
C                 THE NPTS VALUES OF THE DEPENDENT VARIABLE.
C          PARAM  REAL(K), K.GE.NP
C                 THE INITIAL ESTIMATES OF THE NP PARAMETERS.
C          ALPHA  REAL
C                 THE CONVERGENCE CRITERION.  IF THE ABSOLUTE VALUE OF THE
C                 RATIO OF THE PARAMETER CORRECTION TO THE PARAMETER VALUE
C                 DOES NOT EXCEED ALPHA FOR ALL PARAMETERS THEN CONVERGENCE
C                 IS CONSIDERED TO HAVE OCCURRED.  A VALUE OF 1.0E-6 IS
C                 SUGGESTED FOR ALPHA.
C          NITS   INTEGER
C                 THE MAXIMUM NUMBER OF ITERATIONS ALLOWED. A VALUE OF 30
C                 IS SUGGESTED FOR NITS.
C          NPTS   INTEGER
C                 THE NUMBER OF OBSERVATIONS.
C          NP     INTEGER
C                 THE NUMBER OF PARAMETERS.
C          ND1    INTEGER
C                 THE LEADING DIMENSION OF THE MATRIX VARC (SEE EXIT LIST BELOW).
C          LUNOUT INTEGER
C                 THE LOGICAL UNIT NUMBER TO WHICH ERROR MESSAGES AND RESULTS
C                 ARE TO BE PRINTED.
C          JOB    INTEGER
C                 PRINT OPTIONS.  JOB HAS THE DECIMAL EXPANSION ABCDE, WITH
C                 THE FOLLOWING MEANING -
C                     IF A.NE.0, PRINT PARAMETER VALUES AT EACH ITERATION,
C                     IF B.NE.0, PRINT A TABLE OF PARAMETER VALUES, S.E.'S AND
C                         T-VALUES AT CONVERGENCE,
C                     IF C.NE.0, PRINT THE PARAMETER VARIANCE-COVARIANCE MATRIX
C                         AT CONVERGENCE,
C                     IF D.NE.0, PRINT PARAMETER CORRELATIONS AT CONVERGENCE,
C                     IF E.NE.0, PRINT A TABLE OF Y VALUES, FITTED VALUES, AND
C                         RESIDUALS AT CONVERGENCE.
C                 ALL OUTPUT IS WRITTEN TO LOGICAL UNIT LUNOUT.
C          EVAL   SUBROUTINE
C                 A USER-SUPPLIED SUBROUTINE WHICH IS CALLED WITHIN SOLVE
C                 TO EVALUATE THE PREDICTED VALUE FOR THE MODEL AT THE DATA
C                 POINT I AND THE FIRST-ORDER PARTIAL DERIVATIVES WITH
C                 RESPECT TO EACH PARAMETER AT THE DATA POINT I (SEE
C                 SEPARATE DESCRIPTION OF SUBROUTINE EVAL IN EXAMPLE
C                 PROGRAMS).  THE NAME OF THE SUBROUTINE PASSED INTO
C                 SOLVE AS THE ARGUMENT EVAL SHOULD APPEAR IN AN EXTERNAL
C                 STATEMENT IN THE USER'S ROUTINE THAT CALLS SOLVE.
C
C       ON EXIT
C          PARAM  REAL(K), K.GE.NP
C                 THE VALUES OF THE PARAMETERS AT CONVERGENCE OR AT THE FINAL
C                 ITERATION IF CONVERGENCE WAS NOT ACHIEVED.
C          VARC   REAL(ND1,K), ND1.GE.NP, K.GE.NP
```

```
C               THE PARAMETER VARIANCE-COVARIANCE MATRIX.
C       RSS     REAL
C               THE RESIDUAL SUM OF SQUARES FROM THE LAST ITERATION.
C       IFAIL   INTEGER
C               ERROR INDICATOR.
C               IF IFAIL.EQ.0, NO ERRORS WERE DETECTED.
C               IF IFAIL.EQ.1, CONVERGENCE WAS NOT ACHIEVED WITHIN NITS
C                              ITERATIONS.
C               IF IFAIL.EQ.2, THE CALCULATED FIRST DERIVATIVE MATRIX WAS
C                              SINGULAR (THE USER SHOULD CHECK THE DERIVATIVE
C                              CODING IN SUBROUTINE EVAL).
C               IF IFAIL.EQ.8, A USER-FLAGGED ERROR OCCURRED IN SUBROUTINE
C                              EVAL.
C               IF IFAIL.EQ.9, THE NUMBER OF POINTS WAS EQUAL TO THE NUMBER
C                              OF PARAMETERS.
C
C   WORKING STORAGE
C       DERIVS  REAL(K), K.GE.NPTS*NP
C               USED INITIALLY TO STORE THE FIRST-ORDER PARTIAL DERIVATIVES
C               AND LATER TO STORE COMPONENTS OF THE QR DECOMPOSITION.
C       QRAUX   REAL(K), K.GE.NP
C               USED TO STORE INFORMATION FROM THE QR DECOMPOSITION.
C       JPVT    INTEGER(K), K.GE.NP
C               USED TO RECORD COLUMN INTERCHANGES DURING THE QR DECOMPOSITION.
C       DIFF    REAL(K), K.GE.NPTS
C               USED TO STORE THE VECTOR OF DIFFERENCES BETWEEN ACTUAL AND
C               FITTED VALUES.
C       WK1     REAL(K), K.GE.NP*(NP+1)/2
C               A GENERAL WORKING VECTOR.
C
C   SUBROUTINES AND FUNCTIONS REQUIRED
C       USER-SUPPLIED   EVAL
C       NONLIN          DSPLAY
C       LINPACK         SQRDC,SQRSL,STRDI,SAXPY,SDOT,SSCAL,SSWAP,SNRM2,SCOPY
C       FORTRAN         FLOAT,ABS,MOD,AMAX1,MIN0,SQRT,SIGN
C
C   HISTORY
C       14/04/82  PROGRAMMED (R.K. LOWRY)
C
C-------------------------------------------------------------------------
C
      EXTERNAL EVAL
      INTEGER JPVT(1),IFAIL,NPTS,NP,NITS,ND1,LUNOUT,JOB,
     +        I,J,K,L,ITER,ITASK
      REAL DERIVS(NPTS,1),VARC(ND1,1),Y(1),PARAM(1),DIFF(1),
     +     WK1(1),QRAUX(1),DUM(1),
     +     ALPHA,RSS,RES,TEST,RESVAR,ACCUM,DENOM,ABS,FLOAT,ZERO,ONE
      LOGICAL CNVRGD,PRNPAR
C
      DATA ZERO/0.0E0/,ONE/1.0E0/
C
C     INITIALIZE VARIABLES.
      IFAIL=0
      PRNPAR=JOB/10000.NE.0
C
      IF(.NOT.PRNPAR) GO TO 1111
C         PRINT A HEADING IF PARAMETER VALUES ARE REQUIRED AT EACH ITERATION.
          WRITE(LUNOUT,10) (I,I=1,NP)
C         DETERMINE STARTING VALUES AND PRINT THEM.
          RSS=ZERO
          DO 100 I=1,NPTS
```

```
               ITASK=0
               CALL EVAL(PARAM,I,ITASK,WK1,IFAIL)
               IF(IFAIL.NE.0) GO TO 6666
               RES=Y(I)-WK1(1)
               RSS=RSS+RES*RES
  100     CONTINUE
          WRITE(LUNOUT,20) RSS,(PARAM(I),I=1,NP)
C
 1111 CONTINUE
      ITER=1
C
C     REPEAT LOOP TERMINATING WHEN
C              ITERATIONS EXCEED NITS,
C              CONVERGENCE OCCURS,
C          OR A SINGULAR MATRIX IS DETECTED.
C
 2222 CONTINUE
C         CALCULATE A DIFFERENCE VECTOR AND THE MATRIX OF FIRST
C         DERIVATIVES.
          DO 200 I=1,NPTS
               ITASK=0
               CALL EVAL(PARAM,I,ITASK,WK1,IFAIL)
               IF(IFAIL.NE.0) GO TO 6666
               DIFF(I)=Y(I)-WK1(1)
               ITASK=1
               CALL EVAL(PARAM,I,ITASK,WK1,IFAIL)
               IF(IFAIL.NE.0) GO TO 6666
               DO 300 J=1,NP
                  DERIVS(I,J)=WK1(J)
  300          CONTINUE
  200     CONTINUE
C
C         ZERO JPVT TO ALLOW COLUMN INTERCHANGE FOR ALL COLUMNS.
          DO 400 I=1,NP
               JPVT(I)=0
  400     CONTINUE
C
C         SOLVE USING LINPACK ROUTINES.  AFTER THE CALL TO SQRSL WK1 CONTAINS
C         THE CORRECTIONS TO THE PARAMETER ESTIMATES.
C
          CALL SQRDC(DERIVS,NPTS,NPTS,NP,QRAUX,JPVT,WK1,1)
          CALL SQRSL(DERIVS,NPTS,NPTS,NP,QRAUX,DIFF,DUM,DIFF,WK1,DUM,
     +               DUM,100,IFAIL)
C
          IF(IFAIL.NE.0) GO TO 5555
C         CORRECT PARAMETER VALUES AND CHECK FOR CONVERGENCE.
          CNVRGD=.TRUE.
          DO 500 I=1,NP
               J=JPVT(I)
               PARAM(J)=PARAM(J)+WK1(I)
               DENOM=PARAM(J)
               IF(DENOM.EQ.ZERO) DENOM=ONE
               TEST=WK1(I)/DENOM
               IF(ABS(TEST).GT.ALPHA) CNVRGD=.FALSE.
  500     CONTINUE
C         DETERMINE RESIDUAL SUM OF SQUARES.
          RSS=ZERO
          DO 600 I=1,NPTS
               ITASK=0
               CALL EVAL(PARAM,I,ITASK,WK1,IFAIL)
               IF(IFAIL.NE.0) GO TO 6666
```

```
                  RES=Y(I)-WK1(1)
                  RSS=RSS+RES*RES
  600      CONTINUE
C
           IF(.NOT.PRNPAR) GO TO 3333
C              PRINT RESIDUAL SUM OF SQUARES AND PARAMETER ESTIMATES.
               WRITE(LUNOUT,20) RSS,(PARAM(I),I=1,NP)
C
 3333      CONTINUE
C
C          END OF REPEAT LOOP STARTING AT 2222.  TEST FOR COMPLETION.
C
           ITER=ITER+1
           IF(ITER.GT.NITS) GO TO 4444
           IF(.NOT.CNVRGD) GO TO 2222
C
C      NORMAL TERMINATION, CALCULATE VARIANCE-COVARIANCE MATRIX FOR
C      PARAMETERS.
C
       IF(NPTS.EQ.NP) GO TO 7777
       CALL STRDI(DERIVS,NPTS,NP,DUM,O11,IFAIL)
       IF(IFAIL.NE.0) GO TO 5555
       RESVAR=RSS/FLOAT(NPTS-NP)
C
       DO 700 I=1,NP
           DO 800 J=I,NP
               ACCUM=ZERO
               DO 900 K=J,NP
                   ACCUM=ACCUM+DERIVS(I,K)*DERIVS(J,K)
  900          CONTINUE
               ACCUM=ACCUM*RESVAR
               K=JPVT(I)
               L=JPVT(J)
               VARC(K,L)=ACCUM
               VARC(L,K)=ACCUM
  800      CONTINUE
  700 CONTINUE
C      PRINT THE RESULTS.
       CALL DSPLAY(Y,VARC,PARAM,WK1,RESVAR,EVAL,NPTS,NP,ND1,
      +            LUNOUT,JOB,IFAIL)
       IF(IFAIL.NE.0) GO TO 6666
       RETURN
C
 4444 CONTINUE
C      NO CONVERGENCE.
       WRITE(LUNOUT,30) NITS
       IFAIL=1
       RETURN
C
 5555 CONTINUE
C      SINGULAR MATRIX.
       WRITE(LUNOUT,40)
       IFAIL=2
       RETURN
C
 6666 CONTINUE
C      USER-FLAGGED ERROR IN EVAL.
       WRITE(LUNOUT,50)
       IFAIL=8
       RETURN
C
```

```
7777 CONTINUE
C      ZERO RESIDUAL VARIANCE.
       WRITE(LUNOUT,60)
       IFAIL=9
       RETURN
C
C    FORMAT STATEMENTS.
C
    10 FORMAT(//1H ,13X,19HPARAMETER ESTIMATES
      +/1H ,10X,3HRSS,9(I13)/(1H ,13X,9(I13)))
    20 FORMAT(1H ,10(E13.6)/(1H ,13X,9(E13.6)))
    30 FORMAT(//1H ,38H*** SUBROUTINE SOLVE   CONVERGENCE NOT,
      +15H ACHIEVED AFTER,I10,2X,10HITERATIONS//)
    40 FORMAT(//1H ,38H*** SUBROUTINE SOLVE   SINGULAR MATRIX//)
    50 FORMAT(//1H ,41H*** SUBROUTINE SOLVE   USER-FLAGGED ERROR,
      +19H IN SUBROUTINE EVAL//)
    60 FORMAT(//1H ,45H*** SUBROUTINE SOLVE   RESIDUAL VARIANCE ZERO//)
       END


       SUBROUTINE BATES(PARAM,VAR,NITS,NPTS,NP,LUNOUT,JOB,EVAL,BIAS,
      +                 IFAIL,A,DERIVS,RINV,QRAUX,WK1,WK2,JPVT)
C
C----------------------------------------------------------------------
C
C      SUBROUTINE BATES CALCULATES THE BATES AND WATTS MEASURES OF INTRINSIC
C      AND PARAMETER-EFFECTS CURVATURE (BATES AND WATTS, 1980) AND BOX'S
C      MEASURE OF BIAS IN THE PARAMETER ESTIMATORS (BOX, 1971).
C
C      DESCRIPTION
C      SUBROUTINE BATES REPEATEDLY CALLS THE USER-SUPPLIED SUBROUTINE EVAL
C      TO BUILD A MATRIX OF FIRST-ORDER PARTIAL DERIVATIVES AND A MATRIX
C      OF SECOND-ORDER PARTIAL DERIVATIVES, BOTH BASED ON THE PARAMETER
C      ESTIMATES SUPPLIED.  EACH MATRIX IS SCALED BY THE STANDARD RADIUS
C      (BATES AND WATTS, 1980) AND A QR DECOMPOSITION (BUSINGER AND GOLUB,
C      1965) IS APPLIED TO THE FIRST DERIVATIVE MATRIX USING THE LINPACK
C      ROUTINE SQRDC.  THE RESULTS OF THIS DECOMPOSITION ARE THEN USED TO
C      TRANSFORM THE MATRIX OF SECOND DERIVATIVES INTO THE INTRINSIC AND
C      PARAMETER-EFFECTS PORTIONS OF THE ACCELERATION ARRAY, FROM WHICH
C      THE MAXIMUM CURVATURE MEASURES ARE CALCULATED BY AN ITERATIVE
C      PROCEDURE.  THE PARAMETER-EFFECTS PORTION OF THE ACCELERATION ARRAY
C      IS ALSO USED TO OBTAIN BOX'S BIAS MEASURES.  SUBROUTINE WATTS IS
C      CALLED TO PRINT RESULTS AS SPECIFIED BY THE ENTRY ARGUMENT JOB.
C
C      ON ENTRY
C         (NOTE - ALL ARGUMENTS IN THE ENTRY LIST ARE RETURNED UNALTERED.)
C         PARAM   REAL(K), K.GE.NP
C                 THE PARAMETER ESTIMATES FOR THE MODEL AT CONVERGENCE.
C         VAR     REAL
C                 THE RESIDUAL VARIANCE FOR THE MODEL AT CONVERGENCE.
C         NITS    INTEGER
C                 THE MAXIMUM NUMBER OF ITERATIONS ALLOWED FOR THE
C                 DETERMINATION OF MAXIMUM CURVATURE.  A VALUE OF 30 IS
C                 SUGGESTED FOR NITS.
C         NPTS    INTEGER
C                 THE NUMBER OF OBSERVATIONS.
C         NP      INTEGER
C                 THE NUMBER OF PARAMETERS.
C         LUNOUT  INTEGER
C                 THE LOGICAL UNIT NUMBER TO WHICH ERROR MESSAGES AND RESULTS
C                 ARE TO BE PRINTED.
```

```
C       JOB     INTEGER
C               PRINT OPTIONS.  JOB HAS THE DECIMAL EXPANSION ABCDE, WITH
C               THE FOLLOWING MEANING -
C                   IF A.NE.0, PRINT THE MATRIX OF FIRST DERIVATIVES.
C                   IF B.NE.0, PRINT THE FIRST NP FACES OF THE SECOND
C                              DERIVATIVE MATRIX.
C                   IF C.NE.0, PRINT THE FIRST NP FACES OF THE ACCELERATION
C                              ARRAY (PARAMETER-EFFECTS PORTION OF THE ARRAY).
C                   IF D.NE.0, PRINT A TABLE OF PARAMETER ESTIMATOR BIASES
C                              AND PERCENTAGE BIASES.
C                   IF E.NE.0, PRINT MAXIMUM INTRINSIC AND PARAMETER-EFFECTS
C                              CURVATURE MEASURES.
C               ALL OUTPUT IS WRITTEN TO LOGICAL UNIT LUNOUT.
C       EVAL    SUBROUTINE
C               A USER-SUPPLIED SUBROUTINE WHICH IS CALLED WITHIN BATES TO
C               EVALUATE THE FIRST- AND THE SECOND-ORDER PARTIAL DERIVATIVES
C               WITH RESPECT TO THE PARAMETERS AT THE DATA POINT I (SEE
C               SEPARATE DESCRIPTION OF SUBROUTINE EVAL IN EXAMPLE
C               PROGRAMS).  THE NAME OF THE SUBROUTINE PASSED INTO
C               BATES AS THE ARGUMENT EVAL SHOULD APPEAR IN AN EXTERNAL
C               STATEMENT IN THE USER'S ROUTINE THAT CALLS BATES.
C
C    ON EXIT
C       BIAS    REAL(K), K.GE.NP
C               BOX'S ESTIMATE OF BIAS FOR EACH OF THE PARAMETERS.
C       IFAIL   INTEGER
C               ERROR INDICATOR
C                   IF IFAIL.EQ.0, NO ERRORS WERE DETECTED.
C                   IF IFAIL.EQ.3, INCORRECT STANDARD RADIUS (OCCURS WHEN
C                                  VAR.LE.0. OR NP.LE.0).
C                   IF IFAIL.EQ.4, THE CALCULATED FIRST DERIVATIVE MATRIX
C                                  WAS SINGULAR (THE USER SHOULD CHECK THE
C                                  DERIVATIVE CODING IN SUBROUTINE EVAL).
C                   IF IFAIL.EQ.5, SPECIFIED NUMBER OF ITERATIONS EXCEEDED
C                                  IN DETERMINING MAXIMUM INTRINSIC CURVATURE.
C                   IF IFAIL.EQ.6, SPECIFIED NUMBER OF ITERATIONS EXCEEDED
C                                  IN DETERMINING MAXIMUM PARAMETER-EFFECTS
C                                  CURVATURE.
C                   IF IFAIL.EQ.8, A USER-FLAGGED ERROR OCCURRED IN
C                                  SUBROUTINE EVAL.
C
C    WORKING STORAGE
C       A       REAL(K), K.GE.NPTS*NP*NP
C               USED INITIALLY TO STORE THE SECOND-ORDER PARTIAL DERIVATIVES
C               AND LATER THE ACCELERATION ARRAY.
C       DERIVS  REAL(K), K.GE.NPTS*NP
C               USED INITIALLY TO STORE THE FIRST-ORDER PARTIAL DERIVATIVES
C               AND LATER TO STORE COMPONENTS OF THE QR DECOMPOSITION.
C       RINV    REAL(K), K.GE.NP*NP
C               USED TO STORE THE INVERSE OF THE R MATRIX FROM THE QR
C               DECOMPOSITION.  THIS INFORMATION IS USED IN SUBROUTINE
C               ACCEL AND SUBROUTINE BOX.
C       QRAUX   REAL(K), K.GE.NP
C               USED TO STORE INFORMATION FROM THE QR DECOMPOSITION AND
C               AS A TEMPORARY STORAGE ARRAY IN SUBROUTINE CURVE.
C       WK1     REAL(K), K.GE.NP*(NP+1)/2
C               A GENERAL WORKING VECTOR.
C       WK2     REAL(K), K.GE.NPTS
C               A TEMPORARY STORAGE VECTOR USED ONLY IN SUBROUTINE CURVE.
C       JPVT    INTEGER(K), K.GE.NP
C               USED TO RECORD COLUMN INTERCHANGES DURING THE QR DECOMPOSITION.
```

```
C
C       SUBROUTINES AND FUNCTIONS REQUIRED
C          USER-SUPPLIED   EVAL
C          NONLIN          ARRAYS,ACCEL,BOX,CURVE,CNTEST,WATTS
C          LINPACK         SQRDC,SAXPY,SDOT,SSCAL,SSWAP,SNRM2
C          FORTRAN         SIN,COS,ATAN2,ABS,SIGN,MOD,SQRT,AMAX1,MIN0,FLOAT
C
C       HISTORY
C          12/02/81  TRANSLATED AND MODIFIED FROM AN ALGOL PROGRAM
C                                        BY D.M. BATES     (R.K. LOWRY)
C
C-----------------------------------------------------------------------
C       EXTERNAL EVAL
        INTEGER JPVT(1),NITS,NPTS,NP,LUNOUT,JOB,IFAIL,I,J,K,NW,M,ITASK
        REAL PARAM(1),A(NPTS,NP,1),DERIVS(NPTS,1),RINV(NP,1),QRAUX(1),
     +       BIAS(1),WK1(1),WK2(1),VAR,STRAD,TEMP,C1,C2,W,ZERO,SQRT,FLOAT
C
        DATA ZERO/0.0E0/
C
        IFAIL=0
C       CONVERT VARIANCE TO STANDARD RADIUS.
        IF(VAR.LE.ZERO.OR.NP.LE.0) GO TO 2222
        STRAD=SQRT(VAR*FLOAT(NP))
C       CALCULATE FIRST AND SECOND DERIVATIVES, SCALING THEM BY THE
C       STANDARD RADIUS.
        DO 100 I=1,NPTS
           ITASK=1
           CALL EVAL(PARAM,I,ITASK,WK1,IFAIL)
           IF(IFAIL.NE.0) GO TO 4444
           DO 200 J=1,NP
              DERIVS(I,J)=WK1(J)/STRAD
  200      CONTINUE
           ITASK=2
           CALL EVAL(PARAM,I,ITASK,WK1,IFAIL)
           IF(IFAIL.NE.0) GO TO 4444
           NW=0
           DO 300 J=1,NP
              DO 400 K=1,J
                 NW=NW+1
                 W=WK1(NW)/STRAD
                 A(I,J,K)=W
                 IF(J.NE.K) A(I,K,J)=W
  400         CONTINUE
  300      CONTINUE
  100   CONTINUE
C
C       PRINT OUT FIRST- AND SECOND-ORDER DERIVATIVES IF REQUESTED.
C
        CALL ARRAYS(A,DERIVS,STRAD,NPTS,NP,LUNOUT,JOB)
C
C       ZERO JPVT TO ALLOW COLUMN INTERCHANGE FOR ALL COLUMNS.
C
        DO 700 I=1,NP
           JPVT(I)=0
  700   CONTINUE
C
C       PERFORM QR DECOMPOSITION OF FIRST DERIVATIVES.
C
        CALL SQRDC(DERIVS,NPTS,NPTS,NP,QRAUX,JPVT,WK1,1)
C
```

```
C      TRANSFER THE DIAGONAL OF R TO QRAUX AND RESTORE THE LEADING
C      DIAGONAL OF THE HOUSEHOLDER TRANSFORMATIONS TO DERIVS,
C      SCALING THE TRANSFORMATIONS IN THE PROCESS.
C
       DO 500 I=1,NP
          IF(QRAUX(I).EQ.ZERO.OR.DERIVS(I,I).EQ.ZERO) GO TO 3333
          TEMP=QRAUX(I)
          QRAUX(I)=DERIVS(I,I)
          DERIVS(I,I)=TEMP
          DO 600 J=I,NPTS
             DERIVS(J,I)=-DERIVS(J,I)*QRAUX(I)
  600     CONTINUE
  500 CONTINUE
C
C      REDUCE THE SECOND DERIVATIVES TO THE ACCELERATION ARRAY.
C
       CALL ACCEL(A,DERIVS,RINV,QRAUX,WK1,JPVT,NPTS,NP)
C
C      CALCULATE BOX'S BIAS.
C
       CALL BOX(A,RINV,BIAS,WK1,NPTS,NP)
C
C      CALCULATE THE INTRINSIC CURVATURE MEASURES.
C
       M=NP+1
       CALL CURVE(A,DERIVS,WK1,QRAUX,WK2,C1,M,NPTS,NITS,NPTS,NP,IFAIL)
       IF(IFAIL.EQ.0) GO TO 1111
C         REPORT FAILURE TO CONVERGE.
          IFAIL=5
          WRITE(LUNOUT,30) NITS
C
 1111 CONTINUE
C
C      CALCULATE THE PARAMETER-EFFECTS CURVATURE MEASURES.
C
       CALL CURVE(A,DERIVS,WK1,QRAUX,WK2,C2,1,NP,NITS,NPTS,NP,IFAIL)
       IF(IFAIL.EQ.6) WRITE(LUNOUT,30) NITS
C
C      PRINT THE RESULTS.
C
       CALL WATTS(A,BIAS,PARAM,C1,C2,NPTS,NP,LUNOUT,JOB)
C
       RETURN
C
 2222 CONTINUE
C      INCORRECT STANDARD RADIUS.
       IFAIL=3
       WRITE(LUNOUT,10)
       RETURN
C
 3333 CONTINUE
C      SINGULAR MATRIX.
       IFAIL=4
       WRITE(LUNOUT,20)
       RETURN
C
 4444 CONTINUE
C      USER-FLAGGED ERROR IN EVAL.
       IFAIL=8
       WRITE(LUNOUT,40)
       RETURN
```

```
C
C    FORMAT STATEMENTS.
C
     10 FORMAT(//1H ,48H*** SUBROUTINE BATES    INCORRECT STANDARD RADIUS
       +    //)
     20 FORMAT(//1H ,38H*** SUBROUTINE BATES    SINGULAR MATRIX//)
     30 FORMAT(//1H ,38H*** SUBROUTINE BATES    CONVERGENCE NOT,
       +15H ACHIEVED AFTER,I10,2X,10HITERATIONS//)
     40 FORMAT(//1H ,41H*** SUBROUTINE BATES    USER-FLAGGED ERROR,
       +19H IN SUBROUTINE EVAL//)
        END



        SUBROUTINE GNRATE(PARAM,SD,NPTS,LUNOUT,EVAL,Y,IFAIL)
C
C-----------------------------------------------------------------------
C
C    SUBROUTINE GNRATE RETURNS A VECTOR OF SIMULATED DEPENDENT VARIABLE VALUES
C    WITH INDEPENDENT AND IDENTICALLY DISTRIBUTED NORMAL ERROR COMPONENTS.
C
C    DESCRIPTION
C    GNRATE CALLS THE USER-SUPPLIED SUBROUTINE EVAL NPTS TIMES TO GENERATE
C    MODEL VALUES WITH IIDN ERROR COMPONENTS.  THE FUNCTION RANF CALLED IN
C    THE SUBROUTINE NORMAL IS A NON-ANSI FORTRAN ROUTINE WHICH GENERATES
C    UNIFORM PSEUDO-RANDOM NUMBERS.  THE USE OF THIS FUNCTION REQUIRES AN
C    INITIAL CALL TO THE SUBROUTINE RANSET PRIOR TO THE FIRST CALL TO GNRATE
C    TO PROVIDE A SEED FOR RANDOM NUMBER GENERATION.
C
C    ON ENTRY
C       (NOTE - ALL ARGUMENTS IN THE ENTRY LIST ARE RETURNED UNALTERED.)
C       PARAM  REAL(K), K.GE.NP
C              THE VALUES OF THE NP PARAMETERS IN THE MODEL.
C       SD     REAL
C              THE ERROR COMPONENTS ARE TO BE GENERATED FROM A POPULATION
C              WITH A STANDARD DEVIATION OF SD.
C       NPTS   INTEGER
C              THE NUMBER OF POINTS TO BE GENERATED.
C       LUNOUT INTEGER
C              THE LOGICAL UNIT NUMBER TO WHICH ERROR MESSAGES ARE
C              TO BE PRINTED.
C       EVAL   SUBROUTINE
C              A USER-SUPPLIED SUBROUTINE WHICH IS CALLED WITHIN GNRATE TO
C              CALCULATE THE PREDICTED VALUE FOR THE MODEL AT THE DATA
C              POINT I (SEE SEPARATE DESCRIPTION OF SUBROUTINE EVAL IN
C              EXAMPLE PROGRAMS).  THE NAME OF THE SUBROUTINE PASSED INTO
C              GNRATE AS THE ARGUMENT EVAL SHOULD APPEAR IN AN EXTERNAL
C              STATEMENT IN THE USER'S ROUTINE THAT CALLS GNRATE.
C
C    ON EXIT
C       Y      REAL(K), K.GE.NPTS
C              THE GENERATED DEPENDENT VARIABLE WITH IIDN ERRORS.
C       IFAIL  INTEGER
C              ERROR INDICATOR.
C                 IF IFAIL.EQ.0, NO ERRORS WERE DETECTED.
C                 IF IFAIL.EQ.8, A USER-FLAGGED ERROR OCCURRED IN SUBROUTINE
C                    EVAL.
C
C    SUBROUTINES AND FUNCTIONS REQUIRED
C       USER-SUPPLIED          EVAL
C       NONLIN                 NORMAL
C       NON-ANSI FORTRAN       RANF  (A UNIFORM PSEUDO-RANDOM NUMBER GENERATOR)
```

```
C         FORTRAN              SQRT,ALOG
C
C     HISTORY
C        27/04/82  PROGRAMMED (R.K. LOWRY)
C
C--------------------------------------------------------------------------
C     EXTERNAL EVAL
      INTEGER NPTS,N,I,LUNOUT,ITASK,IFAIL
      REAL Y(1),PARAM(1),ERR(2),SD,W
C
      IFAIL=0
C
C     USE THE 2 NORMALLY DISTRIBUTED DEVIATES GENERATED BY SUBROUTINE
C     NORMAL IN TURN UNTIL ALL NPTS DATA POINTS HAVE BEEN GENERATED.
C
      N=0
 1111 CONTINUE
      CALL NORMAL(ERR)
C
      DO 100 I=1,2
         N=N+1
         IF(N.GT.NPTS) GO TO 2222
         ITASK=0
         CALL EVAL(PARAM,N,ITASK,W,IFAIL)
         IF(IFAIL.NE.0) GO TO 3333
         Y(N)=W+ERR(I)*SD
  100 CONTINUE
      GO TO 1111
 2222 CONTINUE
      RETURN
C
 3333 CONTINUE
C     USER-FLAGGED ERROR IN EVAL.
      IFAIL=8
      WRITE(LUNOUT,10)
      RETURN
C
C   FORMAT STATEMENT.
C
   10 FORMAT(//1H ,42H*** SUBROUTINE GNRATE   USER-FLAGGED ERROR,
     +19H IN SUBROUTINE EVAL//)
      END



      SUBROUTINE SUMRY(STMATE,NSIM,LUNOUT,JOB,IFAIL)
C
C--------------------------------------------------------------------------
C
C     SUBROUTINE SUMRY SUPPLIES VARIOUS SUMMARY STATISTICS FOR A VECTOR
C     OF PARAMETER ESTIMATES OBTAINED THROUGH SIMULATION.
C
C     DESCRIPTION
C     SUMRY ORDERS NSIM PARAMETER ESTIMATES PASSED IN AS THE ONE-DIMENSIONAL
C     ARRAY STMATE AND CALCULATES THE MEAN, VARIANCE, SKEWNESS, AND EXCESS
C     KURTOSIS.  THE PRINT OPTION JOB ALLOWS THE USER TO SELECTIVELY PRINT
C     THE VECTOR OF ORDERED VALUES, A TABLE OF MOMENTS, AND A TABLE OF
C     FREQUENCIES.
C
C     ON INPUT
C        (NOTE - ALL ARGUMENTS IN THE ENTRY LIST ARE RETURNED UNALTERED
C        UNLESS DETAILED IN THE EXIT LIST.)
```

```
C          STMATE REAL(K), K.GE.NSIM
C                  A VECTOR OF PARAMETER ESTIMATES GENERATED BY NSIM CALLS TO
C                  GNRATE AND SOLVE.
C          NSIM   INTEGER
C                  THE NUMBER OF VALUES STORED IN STMATE.
C          LUNOUT INTEGER
C                  THE LOGICAL UNIT NUMBER TO WHICH ALL OUTPUT IS TO BE WRITTEN.
C          JOB    INTEGER
C                  PRINT OPTIONS.  JOB HAS THE DECIMAL EXPANSION ABC, WITH THE
C                  FOLLOWING MEANING -
C                     A.NE.0, PRINT THE ORDERED PARAMETER ESTIMATES.
C                     B.NE.0, PRINT THE MOMENTS FOR THE DISTRIBUTION.
C                     C.NE.0, PRINT A FREQUENCY TABLE FOR THE ESTIMATES.
C                  ALL OUTPUT IS WRITTEN TO LOGICAL UNIT LUNOUT.
C
C     ON EXIT
C          STMATE REAL(K), K.GE.NSIM
C                  THE PARAMETER ESTIMATES ORDERED FROM SMALLEST TO LARGEST.
C          IFAIL  INTEGER
C                  ERROR INDICATOR
C                     IF IFAIL.EQ.0, NO ERRORS WERE DETECTED.
C                     IF IFAIL.EQ.7, INSUFFICIENT NUMBER OF VALUES WERE PASSED
C                                    IN STMATE FOR CALCULATION OF MOMENTS.
C
C     SUBROUTINES AND FUNCTIONS REQUIRED
C          NONLIN    SORT,MOMENT,FREQNC
C          FORTRAN   AINT,SQRT,FLOAT,MOD
C
C     HISTORY
C          26/04/82  PROGRAMMED (R.K. LOWRY)
C
C-----------------------------------------------------------------------
C
      INTEGER NSIM,LUNOUT,JOB,MOD,I,IFAIL
      REAL STMATE(1),MEAN,VAR,G1,G2
      LOGICAL PRNVAL,PRNMOM,PRNFRQ
C
C     DETERMINE PRINT OPTIONS.
C
      PRNVAL=JOB/100.NE.0
      PRNMOM=MOD(JOB,100)/10.NE.0
      PRNFRQ=MOD(JOB,10).NE.0
C
C     SORT THE VALUES INTO ASCENDING ORDER.
C
      CALL SORT(STMATE,1,NSIM)
C
      IF(.NOT.PRNVAL) GO TO 1111
C          PRINT THE ORDERED VALUES.
           WRITE(LUNOUT,10) (STMATE(I),I=1,NSIM)
 1111 CONTINUE
      IF(NSIM.LE.1) GO TO 4444
C
C     CALCULATE THE MOMENTS FOR THE DISTRIBUTION.
C
      CALL MOMENT(STMATE,NSIM,MEAN,VAR,G1,G2)
      IF(.NOT.PRNMOM) GO TO 2222
C          PRINT THE MOMENTS.
           WRITE(LUNOUT,20)
```

```
            WRITE(LUNOUT,30) MEAN,VAR
            WRITE(LUNOUT,40) G1,G2
 2222 CONTINUE
      IF(.NOT.PRNFRQ) GO TO 3333
C         PRINT A TABLE OF FREQUENCIES.
          CALL FREQNC(STMATE,NSIM,MEAN,VAR,LUNOUT)
 3333 CONTINUE
      RETURN
C
 4444 CONTINUE
C     TOO FEW DEGREES OF FREEDOM.
      IFAIL=7
      WRITE(LUNOUT,50)
      RETURN
C
C   FORMAT STATEMENTS.
C
   10 FORMAT(////1H ,10X,14HORDERED VALUES//(1H ,10F12.6))
   20 FORMAT(///1H ,10X,23HMOMENTS OF DISTRIBUTION)
   30 FORMAT(/1H ,10X,4HMEAN,17X,F20.9
      +       //1H ,10X,8HVARIANCE,13X,E20.5)
   40 FORMAT(/1H ,10X,20HSKEWNESS COEFFICIENT,1X,F20.9
      +       //1H ,10X,15HEXCESS KURTOSIS,6X,F20.9)
   50 FORMAT(//1H ,30H*** SUBROUTINE SUMRY   TOO FEW,
      +       19H DEGREES OF FREEDOM//)
      END
```

3. Auxiliary Subroutines

```
      SUBROUTINE ACCEL(A,DERIVS,RINV,ALPHA,TEMPV,JPVT,NPTS,NP)
C
C------------------------------------------------------------------------
C
C     SUBROUTINE ACCEL TRANSFORMS THE ARRAY OF SECOND DERIVATIVES TO THE
C     ACCELERATION ARRAY USING THE RESULTS OF THE QR DECOMPOSITION OF THE
C     FIRST DERIVATIVE ARRAY PERFORMED IN SUBROUTINE BATES.
C
C     HISTORY
C         14/04/82  MODIFIED FROM AN ALGOL PROGRAM BY D.M. BATES (R.K. LOWRY)
C
C------------------------------------------------------------------------
C
      INTEGER JPVT(1),NPTS,NP,I,J,K,L,IMIN1,JJ,JPLUS1
      REAL A(NPTS,NP,1),DERIVS(NPTS,1),RINV(NP,1),ALPHA(1),
      +     TEMPV(1),ACCUM,GAMMA,ZERO,ONE
C
      DATA ZERO/0.0E0/,ONE/1.0E0/
C
C     MULTIPLY SECOND DERIVATIVES BY Q TRANSPOSE.
C
      DO 100 I=1,NP
         DO 200 J=1,I
            DO 300 K=1,NP
               ACCUM=ZERO
               DO 400 L=K,NPTS
                  ACCUM=ACCUM+DERIVS(L,K)*A(L,I,J)
  400          CONTINUE
```

```
                          GAMMA=ACCUM/(ALPHA(K)*DERIVS(K,K))
                          DO 500 L=K,NPTS
                             A(L,I,J)=A(L,I,J)+GAMMA*DERIVS(L,K)
   500                 CONTINUE
   300             CONTINUE
                   DO 600 K=1,NPTS
                      A(K,J,I)=A(K,I,J)
   600             CONTINUE
   200         CONTINUE
   100 CONTINUE
C
C     CALCULATE THE INVERSE OF R.
C
       DO 700 I=1,NP
           DO 800 J=1,NP
               TEMPV(J)=ZERO
   800     CONTINUE
           TEMPV(I)=ONE/ALPHA(I)
           IMIN1=I-1
           IF(IMIN1.LT.1) GO TO 1111
               DO 900 JJ=1,IMIN1
                   J=IMIN1-JJ+1
                   JPLUS1=J+1
                   ACCUM=ZERO
                   DO 1000 K=JPLUS1,NP
                      ACCUM=ACCUM+DERIVS(J,K)*TEMPV(K)
  1000             CONTINUE
                   TEMPV(J)=-ACCUM/ALPHA(J)
   900         CONTINUE
  1111     CONTINUE
           DO 1100 J=1,NP
               K=JPVT(J)
               RINV(K,I)=TEMPV(J)
  1100     CONTINUE
   700 CONTINUE
C
C     PERFORM PRE- AND POST-MULTIPLICATION OF A BY RINV TRANSPOSE
C     AND RINV.
C
       DO 1200 I=1,NPTS
           DO 1300 J=1,NP
               DO 1400 K=1,NP
                   ACCUM=ZERO
                   DO 1500 L=1,NP
                      ACCUM=ACCUM+A(I,J,L)*RINV(L,K)
  1500             CONTINUE
                   DERIVS(J,K)=ACCUM
  1400         CONTINUE
  1300     CONTINUE
           DO 1600 J=1,NP
               DO 1700 K=1,J
                   ACCUM=ZERO
                   DO 1800 L=1,NP
                      ACCUM=ACCUM+RINV(L,J)*DERIVS(L,K)
  1800             CONTINUE
                   A(I,J,K)=ACCUM
                   A(I,K,J)=ACCUM
  1700         CONTINUE
  1600     CONTINUE
  1200 CONTINUE
C
```

```
      RETURN
      END



      SUBROUTINE ARRAYS(A,DERIVS,STRAD,NPTS,NP,LUNOUT,JOB)
C
C-----------------------------------------------------------------------
C
C     SUBROUTINE ARRAYS PRINTS THE ENTIRE FIRST DERIVATIVE MATRIX (DERIVS)
C     AND THE NP FACES OF THE SECOND DERIVATIVE MATRIX (A) IF THESE
C     ARE REQUESTED BY THE CURRENT CONTENTS OF JOB.
C     THIS FACILITY IS INCLUDED PRIMARILY TO ALLOW THE USER TO CHECK
C     THE CORRECTNESS OF THE DERIVATIVE CODING IN THE USER-SUPPLIED
C     SUBROUTINE EVAL.
C
C     HISTORY
C        14/04/82  PROGRAMMED  (R.K. LOWRY)
C
C-----------------------------------------------------------------------
C
      INTEGER NPTS,NP,JOB,LUNOUT,I,J,K,MOD
      REAL A(NPTS,NP,1),DERIVS(NPTS,1),STRAD
      LOGICAL PRND1,PRND2
C
C     DETERMINE THE TWO PRINT OPTIONS.
      PRND1=JOB/10000.NE.0
      PRND2=MOD(JOB,10000)/1000.NE.0
C
      IF((.NOT.PRND1).AND.(.NOT.PRND2)) GO TO 1111
C        PRINT THE STANDARD RADIUS.
         WRITE(LUNOUT,10) STRAD
C
 1111 CONTINUE
      IF(.NOT.PRND1) GO TO 2222
C        PRINT FIRST DERIVATIVES.
         WRITE(LUNOUT,20)
         DO 100 I=1,NPTS
            WRITE(LUNOUT,30) I,(DERIVS(I,J),J=1,NP)
  100    CONTINUE
C
 2222 CONTINUE
      IF(.NOT.PRND2) GO TO 3333
C        PRINT THE SECOND DERIVATIVES.
         WRITE(LUNOUT,40)
         DO 200 I=1,NP
            WRITE(LUNOUT,50) I
            DO 300 J=1,NP
               WRITE(LUNOUT,60) (A(I,J,K),K=1,NP)
  300       CONTINUE
  200    CONTINUE
C
 3333 CONTINUE
      RETURN
C
C   FORMAT STATEMENTS.
C
   10 FORMAT(////1H ,10X,17HSTANDARD RADIUS =,E13.6)
   20 FORMAT(////1H ,10X,31HFIRST-ORDER PARTIAL DERIVATIVES,
     +15H (STANDARDIZED)//1H ,10HDATA POINT)
   30 FORMAT(1H ,I10,10X,8E12.5/(1H ,20X,8E12.5))
```

```
   40 FORMAT(////1H ,10X,32HSECOND-ORDER PARTIAL DERIVATIVES,
     +15H (STANDARDIZED)//1H ,10HDATA POINT)
   50 FORMAT(1H ,I5)
   60 FORMAT(1H ,10X,8E12.5/(/1H ,10X,8E12.5))
      END


         SUBROUTINE BOX(A,RINV,BIAS,TEMPV,NPTS,NP)
C
C-------------------------------------------------------------------------
C
C     SUBROUTINE BOX CALCULATES BOX'S BIAS FOR EACH PARAMETER ESTIMATE
C     AND RETURNS THESE VALUES IN THE ARRAY BIAS.
C
C     HISTORY
C        14/04/82  MODIFIED FROM AN ALGOL PROGRAM BY D.M. BATES (R.K. LOWRY)
C
C-------------------------------------------------------------------------
C
      INTEGER NPTS,NP,I,J
      REAL A(NPTS,NP,1),RINV(NP,1),BIAS(1),TEMPV(1),ZERO,FLOAT
C
      DATA ZERO/0.0E0/
C
      DO 100 I=1,NP
         TEMPV(I)=ZERO
         DO 200 J=1,NP
            TEMPV(I)=TEMPV(I)+A(I,J,J)
  200    CONTINUE
         TEMPV(I)=-TEMPV(I)/FLOAT(2*NP)
  100 CONTINUE
C
      DO 300 I=1,NP
         BIAS(I)=ZERO
         DO 400 J=1,NP
            BIAS(I)=BIAS(I)+RINV(I,J)*TEMPV(J)
  400    CONTINUE
  300 CONTINUE

C
      RETURN
      END



         SUBROUTINE CNTEST(DIR,GRAD,COSA,SINA,THQA,NP,CNVGED)
C
C-------------------------------------------------------------------------
C
C     SUBROUTINE CNTEST PROVIDES A CONVERGENCE TEST FOR THE ITERATIVE
C     DETERMINATION OF MAXIMUM INTRINSIC OR PARAMETER-EFFECTS CURVATURE.
C
C     HISTORY
C        14/04/82  MODIFIED FROM AN ALGOL PROGRAM BY D.M. BATES (R.K. LOWRY)
C        •
C-------------------------------------------------------------------------
C
      INTEGER NP,I
      REAL DIR(1),GRAD(1),COSA,SINA,THQA,EPS,PI,TEST,ABS,SIGN,ATAN2,
     +    SGN,ZERO,ONE,TWO,PSF,SQRT
```

```fortran
      LOGICAL CNVGED
      DATA EPS/1.0E-4/,PI/3.141592654E0/,ZERO/0.0E0/,ONE/1.0E0/,
     +TWO/2.0E0/,PSF/7.5E-1/
C
      CNVGED=.FALSE.
      TEST=ONE-EPS
      IF(ABS(COSA).LE.TEST) GO TO 1111
C
C         CONVERGENCE....
          SGN=SIGN(ONE,COSA)
          DO 100 I=1,NP
              DIR(I)=SGN*GRAD(I)
  100     CONTINUE
          CNVGED=.TRUE.
          RETURN
C
C     NONCONVERGENCE.....
 1111 CONTINUE
      SINA=SQRT(ONE-COSA**2)
      IF(SINA.NE.ONE) GO TO 2222
          THQA=PSF*PI/TWO
          RETURN
C
 2222 CONTINUE
      IF(COSA.GE.ZERO) GO TO 3333
          THQA=PSF*(ATAN2(SINA,COSA)+PI)
          RETURN
C
 3333 CONTINUE
      THQA=PSF*(ATAN2(SINA,COSA))
      RETURN
      END




      SUBROUTINE CURVE(A,TEMPML,DIR,GRAD,TEMPL,CMAX,IM,IN,NITS,NPTS,
     +                 NP,IFAIL)
C
C-----------------------------------------------------------------------
C
C     SUBROUTINE CURVE CALCULATES THE MAXIMUM INTRINSIC AND PARAMETER-
C     EFFECTS CURVATURE MEASURES BY AN ITERATIVE PROCEDURE WHICH TERMINATES
C     WHEN CONVERGENCE OCCURS OR WHEN THE SPECIFIED MAXIMUM NUMBER OF
C     ITERATIONS IS EXCEEDED.  FAILURE TO CONVERGE IS FLAGGED BY THE VALUE
C     6 IN IFAIL.
C
C     HISTORY
C         14/04/82  MODIFIED FROM AN ALGOL PROGRAM BY D.M. BATES (R.K. LOWRY)
C
C-----------------------------------------------------------------------
C
      INTEGER IM,IN,NITS,NPTS,NP,IFAIL,I,J,K,ITER,NPMIN1
      REAL A(NPTS,NP,1),TEMPML(NPTS,1),DIR(1),GRAD(1),TEMPL(1),
     +     CMAX,ACCUM,ALENG,SINA,COSA,SIN,COS,SQRT,THQA,C1,C2,ZERO,ONE
      LOGICAL CNVGED
C
      DATA ZERO/0.0E0/,ONE/1.0E0/
C
C     CALCULATE MAXIMUM CURVATURE.
C
```

```
      NPMIN1=NP-1
      IF(NPMIN1.LE.1) GO TO 1111
          DO 400 I=1,NPMIN1
              DIR(I)=ZERO
  400     CONTINUE
 1111 CONTINUE
      DIR(NP)=ONE
C
      ITER=1
C
C     REPEAT LOOP TERMINATING WHEN
C             ITERATIONS EXCEED NITS,
C          OR CONVERGENCE OCCURS.
C
 2222 CONTINUE
C
          DO 500 I=IM,IN
              DO 600 J=1,NP
                  ACCUM=ZERO
                  DO 700 K=1,NP
                      ACCUM=ACCUM+A(I,J,K)*DIR(K)
  700             CONTINUE
                  TEMPML(I,J)=ACCUM
  600         CONTINUE
  500     CONTINUE
C
          DO 800 I=IM,IN
              ACCUM=ZERO
              DO 900 J=1,NP
                  ACCUM=ACCUM+TEMPML(I,J)*DIR(J)
  900         CONTINUE
              TEMPL(I)=ACCUM
  800     CONTINUE
C
          DO 1000 I=1,NP
              ACCUM=ZERO
              DO 1100 J=IM,IN
                  ACCUM=ACCUM+TEMPML(J,I)*TEMPL(J)
 1100         CONTINUE
              GRAD(I)=ACCUM
 1000     CONTINUE
C
          ACCUM=ZERO
          DO 1200 I=1,NP
              ACCUM=ACCUM+GRAD(I)*GRAD(I)
 1200     CONTINUE
          ALENG=SQRT(ACCUM)
          IF(ALENG.EQ.ZERO) GO TO 4444
              DO 1300 I=1,NP
                  GRAD(I)=GRAD(I)/ALENG
 1300         CONTINUE
          COSA=ZERO
          DO 1400 I=1,NP
              COSA=COSA+DIR(I)*GRAD(I)
 1400     CONTINUE
C             TEST FOR CONVERGENCE.
          CALL CNTEST(DIR,GRAD,COSA,SINA,THQA,NP,CNVGED)
          IF(CNVGED) GO TO 4444
C             TEST FOR EXCESSIVE ITERATIONS.
              ITER=ITER+1
```

```
                    IF(ITER.LE.NITS) GO TO 3333
                    IFAIL=6
                    GO TO 4444
      3333          CONTINUE
                    C2=SIN(THQA)/SINA
                    C1=COS(THQA)-C2*COSA
                    DO 1500 I=1,NP
                    DIR(I)=C1*DIR(I)+C2*GRAD(I)
      1500          CONTINUE
                    GO TO 2222
C
C     END OF REPEAT LOOP STARTING AT 2222.
C
 4444 CONTINUE
      DO 1600 I=IM,IN
         DO 1700 J=1,NP
            ACCUM=ZERO
            DO 1800 K=1,NP
               ACCUM=ACCUM+A(I,J,K)*DIR(K)
 1800       CONTINUE
            TEMPML(I,J)=ACCUM
 1700    CONTINUE
 1600 CONTINUE
C
      DO 1900 I=IM,IN
         ACCUM=ZERO
         DO 2000 J=1,NP
            ACCUM=ACCUM+TEMPML(I,J)*DIR(J)
 2000    CONTINUE
         TEMPL(I)=ACCUM
 1900 CONTINUE
C
      ACCUM=ZERO
      DO 2100 I=IM,IN
         ACCUM=ACCUM+TEMPL(I)*TEMPL(I)
 2100 CONTINUE
      CMAX=SQRT(ACCUM)
      RETURN
      END



      SUBROUTINE DSPLAY(Y,VARC,PARAM,WK1,VAR,EVAL,NPTS,NP,ND1,
     +                  LUNOUT,JOB,IFAIL)
C
C-----------------------------------------------------------------------
C
C     SUBROUTINE DSPLAY PRINTS RESULTS FROM SUBROUTINE SOLVE AS REQUESTED
C     BY THE SETTING OF OPTIONS SPECIFIED BY THE CONTENTS OF JOB.
C
C     HISTORY
C        14/04/82  PROGRAMMED (R.K. LOWRY)
C
C-----------------------------------------------------------------------
C     EXTERNAL EVAL
      INTEGER JOB,MOD,LUNOUT,ND1,NP,I,J,NPTS,IFAIL,ITASK
      REAL Y(1),VARC(ND1,1),PARAM(1),WK1(1),VAR,SQRT,T,SE,
     +     YHAT,RES
      LOGICAL PRNTAB,PRNVCV,PRNCOR,PRNFIT
C     DETERMINE PRINT OPTIONS.
```

```
      PRNTAB=MOD(JOB,10000)/1000.NE.0
      PRNVCV=MOD(JOB,1000)/100.NE.0
      PRNCOR=MOD(JOB,100)/10.NE.0
      PRNFIT=MOD(JOB,10).NE.0
C
      IF(.NOT.PRNTAB) GO TO 1111
C         PRINT A TABLE OF PARAMETER ESTIMATES
          WRITE(LUNOUT,10)
          DO 100 I=1,NP
              SE=SQRT(VARC(I,I))
              T=PARAM(I)/SE
              WRITE(LUNOUT,20) I,PARAM(I),SE,T
  100 CONTINUE
C
 1111 CONTINUE
      IF(.NOT.PRNVCV) GO TO 2222
C         PRINT THE VARIANCE-COVARIANCE MATRIX.
          WRITE(LUNOUT,30)
          DO 200 I=1,NP
              WRITE(LUNOUT,40) I,(VARC(I,J),J=1,I)
  200         CONTINUE
C
 2222 CONTINUE
      IF(.NOT.PRNCOR) GO TO 3333
C         PRINT THE CORRELATION MATRIX.
          WRITE(LUNOUT,50)
          DO 300 I=1,NP
              DO 400 J=1,I
                  WK1(J)=VARC(I,J)/SQRT(VARC(I,I)*VARC(J,J))
  400             CONTINUE
              WRITE(LUNOUT,60) I,(WK1(J),J=1,I)
  300         CONTINUE
C
 3333 CONTINUE
      IF(.NOT.PRNFIT) GO TO 4444
C         PRINT A TABLE OF FITTED VALUES AND RESIDUALS.
          WRITE(LUNOUT,70)
          DO 500 I=1,NPTS
              ITASK=0
              CALL EVAL(PARAM,I,ITASK,WK1,IFAIL)
              IF(IFAIL.NE.0) RETURN
              YHAT=WK1(1)
              RES=Y(I)-YHAT
              WRITE(LUNOUT,80) I,Y(I),YHAT,RES
  500         CONTINUE
C
 4444 CONTINUE
      RETURN
C
C   FORMAT STATEMENTS.
C
   10 FORMAT(////1H ,10X,34HPARAMETER ESTIMATES AT CONVERGENCE/
     +/1H ,9HPARAMETER,4X,8HESTIMATE,11X,2HSE,11X,1HT)
   20 FORMAT(1H ,I5,4X,E12.6,1X,E12.6,2X,F10.2)
   30 FORMAT(////1H ,10X,36HPARAMETER VARIANCE-COVARIANCE MATRIX//)
   40 FORMAT(1H ,I5,8(2X,E12.6)/(1H ,5X,8(2X,E12.6)))
   50 FORMAT(////1H ,10X,22HPARAMETER CORRELATIONS//)
   60 FORMAT(1H ,I5,8(2X,F12.6)/(1H ,5X,8(2X,F12.6)))
   70 FORMAT(////1H ,11X,4HUNIT,14X,1HY,9X,6HFITTED,7X,8HRESIDUAL)
```

```
   80 FORMAT(1H ,I15,3E15.6)
      END



      SUBROUTINE FREQNC(X,N,MEAN,VAR,LUNOUT)
C
C---------------------------------------------------------------------------
C
C     SUBROUTINE FREQNC FORMS AND PRINTS A TABLE OF COUNTS OVER THE
C     RANGE OF X-VALUES STANDARDIZED BY THE STANDARD DEVIATION.
C
C     HISTORY
C         01/03/81  PROGRAMMED  (R.K. LOWRY)
C
C---------------------------------------------------------------------------
C
      INTEGER N,LUNOUT,KOUNT,I
      REAL X(1),MEAN,VAR,SD,CATINC,XX,CAT,AINT,SQRT
C
      DATA CATINC/2.5E-1/
C
      SD=SQRT(VAR)
      KOUNT=0
C
C     FIND THE NEAREST INITIAL INCREMENT.
C
      XX=(X(1)-MEAN)/SD
      CAT=AINT(XX/CATINC)*CATINC
C
C     LOOP FOR N DATA POINTS.
C
      WRITE(LUNOUT,10)
      DO 100 I=1,N
          XX=(X(I)-MEAN)/SD
 1111     CONTINUE
          IF(XX.LE.CAT) GO TO 2222
C             CHECK FOR ENTRIES IN CURRENT CATEGORY.
              IF(KOUNT.EQ.0) GO TO 3333
C                 PRINT CATEGORY TOTAL AND RESET KOUNT.
                  WRITE(LUNOUT,20) CAT,KOUNT
                  KOUNT=0
 3333         CONTINUE
C             INCREMENT TO NEXT CATEGORY.
              CAT=CAT+CATINC
              GO TO 1111
C
 2222     CONTINUE
C         PRESENT VALUE BELONGS TO THIS CATEGORY.
          KOUNT=KOUNT+1
  100 CONTINUE
C
C     WRITE OUT LAST CATEGORY.
      WRITE(LUNOUT,20) CAT,KOUNT
      RETURN
C
C   FORMAT STATEMENTS.
C
   10 FORMAT(////1H ,10X,15HFREQUENCY TABLE/)
   20 FORMAT(1H ,10X,F10.2,I10)
      END
```

```
      SUBROUTINE MOMENT(X,N,M1,VAR,G1,G2)
C
C-----------------------------------------------------------------------
C
C     SUBROUTINE MOMENT CALCULATES THE MEAN, VARIANCE, AND SKEWNESS AND
C     EXCESS KURTOSIS COEFFICIENTS FOR THE N VALUES STORED IN X.
C
C     HISTORY
C         10/05/82  PROGRAMMED  (R.K. LOWRY)
C
C-----------------------------------------------------------------------
C
      INTEGER N,I
      REAL X(1),XX,XN,M1,M2,VAR,M3,M4,G1,G2,ZERO,ONE,THREE,FLOAT,
     +     SQRT
C
      DATA ZERO/0.0E0/,ONE/1.0E0/,THREE/3.0E0/
C
      XN=FLOAT(N)
C
C     FIND THE MEAN OF THE DISTRIBUTION.
C
      M1=ZERO
      DO 100 I=1,N
          M1=M1+X(I)
  100 CONTINUE
      M1=M1/XN
C
C     CALCULATE SECOND, THIRD, AND FOURTH MOMENTS.
C
      M2=ZERO
      M3=ZERO
      M4=ZERO
C
      DO 200 I=1,N
          XX=X(I)-M1
          M2=M2+XX**2
          M3=M3+XX**3
          M4=M4+XX**4
  200 CONTINUE
C
      M2=M2/XN
      M3=M3/XN
      M4=M4/XN
C
C     CALCULATE VARIANCE, SKEWNESS AND EXCESS KURTOSIS COEFFICIENTS.
C
      VAR=M2*XN/(XN-ONE)
      G1=M3/(M2*SQRT(M2))
      G2=M4/(M2**2)-THREE
C
      RETURN
      END



      SUBROUTINE NORMAL(ERR)
C
C-----------------------------------------------------------------------
C
```

```
C     SUBROUTINE NORMAL GENERATES TWO IIDN DEVIATES FROM UNIFORM PSEUDO-
C     RANDOM NUMBERS GENERATED BY THE NON-ANSI FORTRAN FUNCTION RANF.
C     THE METHOD USED IS A MODIFIED BOX-MULLER METHOD.
C
C     HISTORY
C        27/04/82  PROGRAMMED  (R.K. LOWRY)
C
C--------------------------------------------------------------------
C
      REAL ERR(2),E1,E2,E3,DEN,FAC,ZERO,ONE,TWO,RANF,SQRT,ALOG
C
      DATA ZERO/0.0E0/,ONE/1.0E0/,TWO/2.0E0/
C
 1111 CONTINUE
C***
      E1=RANF(0)
C***
      IF(E1.EQ.ZERO) GO TO 1111
C
 2222 CONTINUE
C***
      E2=TWO*RANF(0)-ONE
      E3=TWO*RANF(0)-ONE
C***
C
      DEN=E2*E2+E3*E3
      IF(DEN.GT.ONE) GO TO 2222
C
      FAC=SQRT(-TWO*ALOG(E1))
      ERR(1)=FAC*(E2*E2-E3*E3)/DEN
      ERR(2)=FAC*TWO*E2*E3/DEN
C
      RETURN
      END




      SUBROUTINE SORT(A,II,JJ)
C
C--------------------------------------------------------------------
C
C     SUBROUTINE SORT TAKES THE PORTION OF THE REAL ARRAY A BETWEEN A(II)
C     AND A(JJ) INCLUSIVE AND SORTS IT INTO INCREASING ORDER OF MAGNITUDE.
C     ARRAYS IU(K) AND IL(K) PERMIT SORTING OF UP TO 2**(K+1)-1 ELEMENTS.
C
C     HISTORY
C        01/03/69  PUBLISHED  (R.C. SINGLETON)
C        20/05/82  REQUIREMENT FOR NORMALIZED REALS DROPPED  (R.K. LOWRY)
C
C--------------------------------------------------------------------
C
      INTEGER IU(16),IL(16),I,II,J,JJ,IJ,K,M,L
      REAL A(1),T,TT
C
      M=1
      I=II
      J=JJ
    5 IF(I.GE.J) GO TO 70
   10 K=I
      IJ=(J+I)/2
      T=A(IJ)
```

```
         IF(A(I).LE.T) GO TO 20
         A(IJ)=A(I)
         A(I)=T
         T=A(IJ)
   20 L=J
         IF(A(J).GE.T) GO TO 40
         A(IJ)=A(J)
         A(J)=T
         T=A(IJ)
         IF(A(I).LE.T) GO TO 40
         A(IJ)=A(I)
         A(I)=T
         T=A(IJ)
         GO TO 40
   30 A(L)=A(K)
         A(K)=TT
   40 L=L-1
         IF(A(L).GT.T) GO TO 40
         TT=A(L)
   50 K=K+1
         IF(A(K).LT.T) GO TO 50
         IF(K.LE.L) GO TO 30
         IF(L-I.LE.J-K) GO TO 60
         IL(M)=I
         IU(M)=L
         I=K
         M=M+1
         GO TO 80
   60 IL(M)=K
         IU(M)=J
         J=L
         M=M+1
         GO TO 80
   70 M=M-1
         IF(M.EQ.0) RETURN
         I=IL(M)
         J=IU(M)
   80 IF(J-I.GE.11) GO TO 10
         IF(I.EQ.II) GO TO 5
         I=I-1
   90 I=I+1
         IF(I.EQ.J) GO TO 70
         T=A(I+1)
         IF(A(I).LE.T) GO TO 90
         K=I
  100 A(K+1)=A(K)
         K=K-1
         IF(T.LT.A(K)) GO TO 100
         A(K+1)=T
         GO TO 90
         END



         SUBROUTINE WATTS(A,BIAS,PARAM,C1,C2,NPTS,NP,LUNOUT,JOB)
C
C------------------------------------------------------------------------
C
C        SUBROUTINE WATTS PRINTS RESULTS FROM SUBROUTINE BATES AS REQUESTED
C        BY THE SETTING OF OPTIONS SPECIFIED BY THE CONTENTS OF JOB.
```

```
C
C      HISTORY
C          14/04/82 PROGRAMMED  (R.K. LOWRY)
C
C-----------------------------------------------------------------------
C
       INTEGER NPTS,NP,LUNOUT,JOB,MOD,I,J,K
       REAL A(NPTS,NP,1),BIAS(1),PARAM(1),C1,C2,PERCEN,ZERO,HUNDRD
       LOGICAL PRNC,PRNB,PRNA
C
       DATA ZERO/0.0E0/,HUNDRD/1.0E2/
C
C      DETERMINE PRINT OPTIONS.
C
       PRNC=MOD(JOB,10).NE.0
       PRNB=MOD(JOB,100)/10.NE.0
       PRNA=MOD(JOB,1000)/100.NE.0
C
       IF(.NOT.PRNA) GO TO 1111
C          PRINT ACCELERATION ARRAY.
           WRITE(LUNOUT,10)
           DO 200 I=1,NP
               WRITE(LUNOUT,20)
               DO 300 J=1,NP
                   WRITE(LUNOUT,30) (A(I,J,K),K=1,NP)
  300          CONTINUE
  200      CONTINUE
C
 1111 CONTINUE
       IF(.NOT.PRNB) GO TO 3333
C          PRINT PARAMETER BIASES.
           WRITE(LUNOUT,40)
           DO 400 I=1,NP
               IF(PARAM(I).EQ.ZERO) GO TO 2222
                   PERCEN=BIAS(I)/PARAM(I)*HUNDRD
                   WRITE(LUNOUT,50) I,PARAM(I),BIAS(I),PERCEN
                   GO TO 400
 2222          CONTINUE
               WRITE(LUNOUT,50) I,PARAM(I),BIAS(I)
  400      CONTINUE
C
 3333 CONTINUE
       IF(.NOT.PRNC) GO TO 4444
C          PRINT MAXIMUM CURVATURE MEASURES.
           WRITE(LUNOUT,60) C1,C2
C
 4444 CONTINUE
       RETURN
C
C   FORMAT STATEMENTS.
C
   10 FORMAT(//1H ,10X,18HACCELERATION ARRAY,
      +28H (PARAMETER-EFFECTS PORTION))
   20 FORMAT(/1H )
   30 FORMAT(1H ,10X,10F12.4)
   40 FORMAT(//1H ,10X,10HBOX'S BIAS///1H ,8X,9HPARAMETER,9X,
      +11HLS ESTIMATE,16X,4HBIAS,8X,12HPERCENT BIAS/)
   50 FORMAT(1H ,6X,9HPARAMETER,I2,2E20.8,F20.4)
   60 FORMAT(////1H ,10X,26HMAXIMUM CURVATURE MEASURES//1H ,
```

```
+14HINTRINSIC (IN),11X,F12.4//1H ,22HPARAMETER-EFFECTS (PE),
+3X,F12.4)
END
```

4. Example Programs

```
C     PROGRAM XAMPLE(INPUT,OUTPUT,TAPE5=INPUT,TAPE6=OUTPUT)
C
C----------------------------------------------------------------------
C
C     PROGRAM XAMPLE PROVIDES AN EXAMPLE OF THE USE OF NONLIN SUBROUTINES
C     SOLVE, BATES, GNRATE, AND SUMRY TO OBTAIN PARAMETER ESTIMATES FOR
C     A MODEL/DATA SET COMBINATION AND TO EXAMINE THE PROPERTIES OF
C     THE PARAMETER ESTIMATORS.
C
C     DESCRIPTION
C     AN ASYMPTOTIC REGRESSION MODEL IS USED IN THIS EXAMPLE TO ILLUSTRATE
C     THE USE OF EACH OF THE NONLIN SUBROUTINES.  THE MODEL IS OF THE FORM
C
C          E(Y) = PARAM(1) - PARAM(2) * PARAM(3) ** X ,
C
C     WHERE X IS THE INDEPENDENT VARIABLE, PARAM(1), PARAM(2), AND
C     PARAM(3) ARE THREE PARAMETERS TO BE ESTIMATED FOR A DATA SET,
C     AND E(Y) IS THE EXPECTED VALUE OF THE DEPENDENT VARIABLE, Y.
C     AN ADDITIVE ERROR TERM IS ILLUSTRATED FOR THE FOLLOWING
C     DATA SET (DATA SET 2, APPENDIX 5.A) -
C
C          X    12   23   40   92  156  215
C          Y  .094 .119 .199 .260 .309 .331 .
C
C     THE INITIAL CALL TO SUBROUTINE SOLVE FINDS THE LEAST-SQUARES
C     PARAMETER ESTIMATES FOR THIS DATA SET.  SUBROUTINE BATES IS THEN
C     CALLED TO ESTIMATE THE BIAS IN EACH PARAMETER ESTIMATOR AS WELL AS
C     THE INTRINSIC AND PARAMETER-EFFECTS CURVATURE MEASURES.  THE FINAL
C     STEP IN THE PROGRAM INVOLVES THE USE OF SIMULATION TO STUDY THE
C     SAMPLING PROPERTIES OF THE PARAMETER ESTIMATORS.  SUBROUTINE GNRATE
C     IS CALLED REPEATEDLY TO GENERATE DATA SETS FROM THE MODEL
C
C          Y = PARAM(1) - PARAM(2) * PARAM(3) ** X + E ,
C
C     WHERE PARAM(1), PARAM(2), AND PARAM(3) ARE THE PARAMETER ESTIMATES
C     OBTAINED FOR THE ORIGINAL DATA SET AND E IS AN IIDN COMPONENT
C     DRAWN AT RANDOM FROM A POPULATION WHOSE VARIANCE IS EQUAL TO THE
C     RESIDUAL VARIANCE OBTAINED FOR THE ORIGINAL DATA SET.  SUBROUTINE
C     SOLVE IS CALLED AFTER EVERY CALL TO SUBROUTINE GNRATE TO ESTIMATE
C     THE PARAMETERS FOR THE SIMULATED DATA SET AND THESE PARAMETER
C     ESTIMATES ARE STORED IN THE ARRAY STMATE.  AFTER A SPECIFIED
C     NUMBER OF DATA SETS (USUALLY 1000) HAVE BEEN SIMULATED, A CALL
C     IS MADE TO SUMRY FOR EACH PARAMETER IN THE MODEL TO PROVIDE VARIOUS
C     SUMMARY STATISTICS FOR THE SET OF PARAMETER ESTIMATES.
C     NOTE THAT MODEL SPECIFICATIONS REQUIRED BY THE NONLIN ROUTINES (THE
C     MODEL FUNCTION AND FIRST- AND SECOND-ORDER PARTIAL DERIVATIVES) MUST
C     BE INCLUDED WITH THE USER'S PROGRAM AS A SUBROUTINE (SEE SEPARATE
C     DESCRIPTION OF SUBROUTINE EVAL).
C
C     SUBROUTINES AND FUNCTIONS REQUIRED
C        USER-SUPPLIED         EVAL
C        NONLIN                SOLVE,BATES,GNRATE,SUMRY,DSPLAY,ACCEL,BOX,
C                              CURVE,CNTEST,WATTS,NORMAL,SORT,MOMENT,FREQNC,
```

```
C                               ARRAYS
C               LINPACK         SQRDC,SQRSL,STRDI,SAXPY,SDOT,SSCAL,SSWAP,
C                               SNRM2,SCOPY
C               NON-ANSI FORTRAN RANSET (PROVIDES A SEED FOR RANDOM NUMBER
C                                       GENERATOR)
C                               RANF   (A PSEUDO-RANDOM NUMBER GENERATOR)
C               FORTRAN         ALOG,FLOAT,ABS,MOD,AMAX1,MINO,SQRT,SIGN,SIN,
C                               COS,ATAN2,AINT
C
C       HISTORY
C           18/06/82  PROGRAMMED  (R.K. LOWRY)
C
C-----------------------------------------------------------------------
C
        EXTERNAL EVAL
C
        INTEGER JPVT(3),LUNIN,LUNOUT,NPTS,NP,NITS,ND1,I,J,JOB,NSIM,NS,
       +        IFAIL
        REAL Y(6),PARAM(3),PAR(3),VARC(3,3),DERIVS(18),A(54),QRAUX(3),
       +     WK1(6),WK2(6),DIFF(6),BIAS(3),STMATE(1000,4),X,ALPHA,SEED,
       +     RSS,VAR,SD,FLOAT,SQRT,ALOG
C
C       X-VALUES ARE PASSED TO SUBROUTINE EVAL IN A COMMON BLOCK.
C
        COMMON X(6)
C
C       INITIAL PARAMETER ESTIMATES.
C
        DATA PARAM(1)/3.4E-1/,PARAM(2)/3.0E-1/,PARAM(3)/9.8E-1/
C
C       VALUES FOR CONSTANTS.
C
        DATA LUNIN/5/,LUNOUT/6/,NPTS/6/,NP/3/,NITS/30/,ND1/3/
        DATA ALPHA/1.0E-6/
C
C       AN INITIAL SEED IS REQUIRED BY THE PSEUDO-RANDOM NUMBER GENERATOR
C       RANF.  THE VARIABLE SEED SHOULD BE INITIALIZED WITH A RANDOMLY
C       CHOSEN REAL NUMBER ON EACH PROGRAM RUN.
C
        DATA SEED/2.962357E0/
C
C       READ AND PRINT THE X- AND Y-VALUES.
C
        WRITE(LUNOUT,10)
        DO 100 I=1,NPTS
            READ(LUNIN,20) X(I),Y(I)
            WRITE(LUNOUT,30) X(I),Y(I)
  100   CONTINUE
C
C  A. FIND THE LEAST-SQUARES PARAMETER ESTIMATES.
C
        JOB=11111
        CALL SOLVE(Y,PARAM,ALPHA,NITS,NPTS,NP,ND1,LUNOUT,JOB,EVAL,
       +           VARC,RSS,IFAIL,DERIVS,QRAUX,JPVT,DIFF,WK1)
        IF(IFAIL.NE.0) STOP
C
C  B. CALCULATE PARAMETER BIASES AND CURVATURE MEASURES.
C
        JOB=111
        VAR=RSS/FLOAT(NPTS-NP)
        CALL BATES(PARAM,VAR,NITS,NPTS,NP,LUNOUT,JOB,EVAL,BIAS,IFAIL,
```

```
     +              A,DERIVS,VARC,QRAUX,WK1,WK2,JPVT)
       IF(IFAIL.NE.0) STOP
C
C  C. SIMULATE NSIM DATA SETS, SOLVING FOR EACH IN TURN.  THE DATA SETS
C     ARE GENERATED USING THE PARAMETER ESTIMATES AND THE RESIDUAL
C     STANDARD DEVIATION OBTAINED FROM THE INITIAL FIT ABOVE.
C
       NSIM=1000
       JOB=1001
       IF(NSIM.GT.10) JOB=0
       SD=SQRT(VAR)
C
C     PROVIDE A SEED FOR THE PSEUDO-RANDOM NUMBER GENERATOR (NOTE THAT
C     THIS IS A NON-ANSI FORTRAN SUBROUTINE).
C***
       CALL RANSET(SEED)
C***
       DO 200 I=1,NSIM
C          MAKE A COPY OF THE ORIGINAL PARAMETER ESTIMATES.
           DO 300 J=1,NP
               PAR(J)=PARAM(J)
  300      CONTINUE
C          GENERATE A DATA SET.
           CALL GNRATE(PAR,SD,NPTS,LUNOUT,EVAL,Y,IFAIL)
C          FIND THE PARAMETER ESTIMATES FOR THE DATA SET.
           CALL SOLVE(Y,PAR,ALPHA,NITS,NPTS,NP,ND1,LUNOUT,JOB,EVAL,
     +                VARC,RSS,IFAIL,DERIVS,QRAUX,JPVT,DIFF,WK1)
           IF(IFAIL.NE.0) STOP
C          SAVE THE PARAMETER ESTIMATES.
           DO 400 J=1,NP
               STMATE(I,J)=PAR(J)
  400      CONTINUE
  200 CONTINUE
C
C     SUMMARIZE THE SIMULATION RESULTS.
C
       JOB=11
       DO 500 I=1,3
           WRITE(LUNOUT,40) I
C          NOTE THAT THE PASSING OF A COLUMN OF A TWO-DIMENSIONAL ARRAY,
C          I.E. STMATE(1,I), TO A ONE-DIMENSIONAL ARRAY IN SUMRY WILL NOT
C          GIVE THE CORRECT RESULTS IF ELEMENTS OF A COLUMN DO NOT OCCUPY
C          CONTIGUOUS WORDS IN MEMORY.
           CALL SUMRY(STMATE(1,I),NSIM,LUNOUT,JOB,IFAIL)
  500 CONTINUE
C
       STOP
C
C     FORMAT STATEMENTS.
C
   10 FORMAT(///1H ,10X,20HLEAF PRODUCTION DATA
     +          //1H ,9X,1HX,9X,1HY//)
   20 FORMAT(F5.1,F5.3)
   30 FORMAT(1H ,5X,F5.1,5X,F5.3)
   40 FORMAT(///1H ,10X,13HPARAMETER NO.,I4)
       END



       SUBROUTINE EVAL(PARAM,I,ITASK,WK1,IFAIL)
C
C-----------------------------------------------------------------------
```

```
C
C          SUBROUTINE EVAL MUST BE SUPPLIED BY THE USER TO ALLOW THE CALCULATION
C          OF PREDICTED VALUES AND THE EVALUATION OF THE FIRST- AND SECOND-ORDER
C          PARTIAL DERIVATIVES FOR A MODEL.
C
C          DESCRIPTION
C          SUBROUTINE EVAL IS CALLED BY SUBROUTINE SOLVE TO CALCULATE PREDICTED
C          VALUES AND FIRST DERIVATIVES REQUIRED IN THE ESTIMATION OF PARAMETERS
C          FOR A MODEL, BY SUBROUTINE BATES TO CALCULATE THE FIRST AND SECOND
C          DERIVATIVES REQUIRED IN THE ESTIMATION OF PARAMETER BIASES AND THE
C          CURVATURE MEASURES, AND BY SUBROUTINE GNRATE TO CALCULATE PREDICTED
C          VALUES REQUIRED IN A SIMULATION STUDY.  THE USER MUST THEREFORE
C          INCLUDE SUBROUTINE EVAL (WITH APPROPRIATE CODE) WITH ANY PROGRAM
C          THAT CALLS ONE OR MORE OF THESE ROUTINES.  THE FORTRAN CODE
C          ACCOMPANYING THIS DESCRIPTION PROVIDES AN EXAMPLE TO FOLLOW WHEN
C          CODING A NEW MODEL.
C
C          ON ENTRY
C              PARAM  REAL(K), K.GE.NP
C                     THE CURRENT PARAMETER ESTIMATES.
C              I      INTEGER
C                     THE INDEX TO THE CURRENT X-VALUE(S) TO BE USED.  EACH CALL
C                     TO EVAL BY A NONLIN ROUTINE EVALUATES EITHER A PREDICTED
C                     VALUE, THE FIRST DERIVATIVES OR THE SECOND DERIVATIVES
C                     AT THE DATA POINT I WITH I BEING GIVEN THE VALUES 1,...,NPTS
C                     SUCCESSIVELY OVER NPTS CALLS.
C              ITASK  INTEGER
C                     ITASK IS SET TO EITHER 0, 1 OR 2 BY A NONLIN CALLING ROUTINE
C                     TO INDICATE A SPECIFIC TASK FOR EACH CALL.
C                     IF ITASK.EQ.0, THE CALCULATION OF A PREDICTED VALUE FOR THE
C                                POINT I IS REQUIRED.
C                     IF ITASK.EQ.1, THE CALCULATION OF FIRST DERIVATIVES FOR THE
C                                POINT I IS REQUIRED.
C                     IF ITASK.EQ.2, THE CALCULATION OF SECOND DERIVATIVES FOR THE
C                                POINT I IS REQUIRED.
C                     AN EXAMPLE OF THE FORTRAN CODE REQUIRED FOR THE CORRECT
C                     USE OF ITASK FOLLOWS THIS DESCRIPTION.
C
C          ON EXIT
C              WK1    REAL(K), K.GE.NP*(NP+1)/2
C                     A VECTOR OF RESULTS FOR RETURN TO THE NONLIN CALLING ROUTINE.
C
C                     WHEN ITASK=0, THE FIRST LOCATION OF WK1 SHOULD CONTAIN THE
C                     PREDICTED VALUE FOR THE MODEL AT THE DATA POINT I.
C
C                     WHEN ITASK=1, WK1 SHOULD CONTAIN NP FIRST DERIVATIVES STORED
C                     IN ORDER OF INCREASING PARAM INDICES.  I.E., IF D(J) IS THE
C                     VALUE OF THE FIRST DERIVATIVE EVALUATED FOR PARAM(J) AT THE
C                     POINT I, THEN THE VECTOR WK1 WILL CONTAIN THE VALUES D(J)
C                     IN THE FOLLOWING ORDER -
C                         WK1(1)=D(1), WK1(2)=D(2), WK1(3)=D(3), ..., WK1(NP)=D(NP).
C
C                     WHEN ITASK=2, WK1 SHOULD CONTAIN NP*(NP+1)/2 SECOND
C                     DERIVATIVES STORED IN SYMMETRIC MODE.  IF D(J,K) IS THE
C                     VALUE OF THE SECOND DERIVATIVE EVALUATED FOR PARAM(J) AND
C                     PARAM(K) AT THE DATA POINT I, THEN THE VECTOR WK1 WILL CONTAIN
C                     THE VALUES D(J,K) IN THE FOLLOWING ORDER -
C                         WK1(1)=D(1,1), WK1(2)=D(2,1), WK1(3)=D(2,2),
C                         WK1(4)=D(3,1), WK1(5)=D(3,2), WK1(6)=D(3,3),
C                         WK1(7)=D(4,1), ..., WK1(NP*(NP+1)/2)=D(NP,NP).
```

```
C         IFAIL   INTEGER
C                 AN ERROR FLAG THAT MAY BE SET TO A NONZERO VALUE IN EVAL TO
C                 INDICATE TO THE NONLIN CALLING ROUTINE THAT AN ERROR HAS
C                 OCCURRED IN SUBROUTINE EVAL.  WHEN THIS OCCURS, THE NONLIN
C                 CALLING ROUTINE IMMEDIATELY TERMINATES WITH AN ERROR MESSAGE.
C
C         HISTORY
C                 28/05/82  PROGRAMMED  (R.K. LOWRY)
C
C----------------------------------------------------------------------
C
C
C
C          EXAMPLE PROGRAM FOR THE MODEL
C
C                          E(Y) = PARAM(1) - PARAM(2) * PARAM(3) ** X
C
C
          INTEGER I,ITASK,IFAIL
          REAL PARAM(3),WK1(6),X,ZERO,ONE,TWO
C
C         A COMMON STATEMENT IS REQUIRED TO ALLOW EVAL ACCESS TO X VALUES.
C         A CORRESPONDING COMMON STATEMENT IS ALSO REQUIRED IN A USER'S ROUTINE
C         IN WHICH X-VALUES ARE ACCESSIBLE.  THE DIMENSION OF X SHOULD BE AT
C         LEAST AS LARGE AS NPTS.  NOTE THAT THE USER-DEFINED COMMON BLOCK
C         ALLOWS THE USER TO SPECIFY MODELS WITH MORE THAN ONE X VARIATE.
C
          COMMON X(6)
C
          DATA ZERO/0.0E0/,ONE/1.0E0/,TWO/2.0E0/
C
C         CLEAR THE ERROR FLAG.
C
          IFAIL=0
C
          IF(ITASK.NE.0) GO TO 1111
C             PREDICTED VALUE IS REQUIRED.
              WK1(1)=PARAM(1)-PARAM(2)*PARAM(3)**X(I)
              RETURN
C
 1111 CONTINUE
          IF(ITASK.NE.1) GO TO 2222
C             FIRST DERIVATIVES REQUIRED.
              WK1(1)=ONE
              WK1(2)=-PARAM(3)**X(I)
              WK1(3)=-PARAM(2)*X(I)*PARAM(3)**(X(I)-ONE)
              RETURN
C
 2222 CONTINUE
          IF(ITASK.NE.2) GO TO 3333
C             SECOND DERIVATIVES REQUIRED.
              WK1(1)=ZERO
              WK1(2)=ZERO
              WK1(3)=ZERO
              WK1(4)=ZERO
              WK1(5)=-X(I)*PARAM(3)**(X(I)-ONE)
              WK1(6)=-PARAM(2)*X(I)*(X(I)-ONE)*PARAM(3)**(X(I)-TWO)
              RETURN
C
 3333 CONTINUE
```

```
C     ERROR - SET IFAIL TO A NONZERO VALUE.
      IFAIL=1
      RETURN
      END
```

LEAF PRODUCTION DATA

X	Y
12.0	.094
23.0	.119
40.0	.199
92.0	.260
156.0	.309
215.0	.331

PARAMETER ESTIMATES

RSS	1	2	3
.268175E-02	.340000E+00	.300000E+00	.980000E+00
.620509E-03	.331851E+00	.292823E+00	.983632E+00
.605641E-03	.334930E+00	.295523E+00	.983811E+00
.605635E-03	.335012E+00	.295514E+00	.983828E+00
.605635E-03	.335020E+00	.295512E+00	.983829E+00
.605635E-03	.335020E+00	.295512E+00	.983830E+00
.605635E-03	.335020E+00	.295512E+00	.983830E+00

PARAMETER ESTIMATES AT CONVERGENCE

PARAMETER	ESTIMATE	SE	T
1	.335020E+00	.173160E-01	19.35
2	.295512E+00	.195638E-01	15.11
3	.983830E+00	.370336E-02	265.66

PARAMETER VARIANCE-COVARIANCE MATRIX

1	.299845E-03		
2	.922989E-04	.382742E-03	
3	.535329E-04	-.136012E-04	.137148E-04

PARAMETER CORRELATIONS

1	1.000000		
2	.272455	1.000000	
3	.834789	-.187729	1.000000

UNIT	Y	FITTED	RESIDUAL
1	.940000E-01	.920159E-01	.198406E-02
2	.119000E+00	.131909E+00	-.129094E-01
3	.199000E+00	.181073E+00	.179269E-01
4	.260000E+00	.269071E+00	-.907140E-02
5	.309000E+00	.311789E+00	-.278862E-02
6	.331000E+00	.326142E+00	.485843E-02

ACCELERATION ARRAY (PARAMETER-EFFECTS PORTION)

.1504	.4049	-.0321
.4049	1.0903	-.0865
-.0321	-.0865	-.0806

-.0445	-.1198	-.0244
-.1198	-.3224	-.0657
-.0244	-.0657	-.0134

.0981	.2641	.0538
.2641	.7110	.1450
.0538	.1450	.0296

BOX'S BIAS

PARAMETER	LS ESTIMATE	BIAS	PERCENT BIAS
PARAMETER 1	.33502050E+00	.29713907E-02	.8869
PARAMETER 2	.29551209E+00	.47364067E-02	1.6028
PARAMETER 3	.98382970E+00	-.21805994E-03	-.0222

MAXIMUM CURVATURE MEASURES

INTRINSIC (IN)	.2318
PARAMETER-EFFECTS (PE)	1.5264

PARAMETER NO. 1
MOMENTS OF DISTRIBUTION

MEAN	.337936083
VARIANCE	.42018E-03

SKEWNESS COEFFICIENT   1.091563990

EXCESS KURTOSIS   2.460111097

FREQUENCY TABLE

-2.25	1
-2.00	1
-1.75	10
-1.50	17
-1.25	36
-1.00	61
-.75	100
-.50	108
-.25	107
0.00	124
.25	97
.50	81
.75	76
1.00	41
1.25	36
1.50	28
1.75	21
2.00	17
2.25	10
2.50	4
2.75	3
3.00	8
3.25	7
3.75	1
4.00	1
4.50	2
4.75	1
5.75	1

PARAMETER NO.   2

MOMENTS OF DISTRIBUTION

MEAN   .300715732

VARIANCE   .41175E-03

SKEWNESS COEFFICIENT   .129410409

EXCESS KURTOSIS   .286503113

FREQUENCY TABLE

-3.00	2
-2.50	5

```
        -2.25          7
        -2.00          8
        -1.75         19
        -1.50         22
        -1.25         33
        -1.00         46
         -.75         70
         -.50         99
         -.25        107
         0.00        103
          .25         88
          .50         96
          .75         77
         1.00         52
         1.25         53
         1.50         45
         1.75         26
         2.00         22
         2.25          9
         2.50          4
         2.75          3
         3.00          2
         4.00          2
```

PARAMETER NO.   3

MOMENTS OF DISTRIBUTION

MEAN   .983545565

VARIANCE   .15189E-04

SKEWNESS COEFFICIENT   -.287230252

EXCESS KURTOSIS   .081067991

FREQUENCY TABLE

```
        -3.50          1
        -3.25          1
    .   -3.00          2
        -2.75          4
        -2.50          6
        -2.25          4
        -2.00         13
        -1.75         22
        -1.50         12
        -1.25         35
        -1.00         44
         -.75         77
         -.50         89
         -.25         89
```

0.00	83
.25	102
.50	89
.75	79
1.00	87
1.25	65
1.50	39
1.75	27
2.00	15
2.25	11
2.50	3
2.75	1

```
C      PROGRAM COMPRE(INPUT,OUTPUT,TAPE5=INPUT,TAPE6=OUTPUT)
C
C--------------------------------------------------------------------------
C
C      PROGRAM COMPRE ILLUSTRATES THE USE OF SUBROUTINE SOLVE IN THE
C      COMPARISON OF PARAMETER ESTIMATES BETWEEN DIFFERENT DATA SETS BY
C      THE EXTRA SUM OF SQUARES PRINCIPLE (DRAPER AND SMITH, 1981).
C
C      DESCRIPTION
C      IN THIS EXAMPLE, FOUR DATA SETS WITH 42, 41, 42, AND 41 DATA POINTS
C      (APPENDIX 3.A) ARE COMBINED TO FORM A DATA SET OF 166 POINTS.  THE
C      MODEL TO BE FITTED TO THESE DATA POINTS IS A TWO-PARAMETER ASYMPTOTIC
C      YIELD-DENSITY MODEL WITH MULTIPLICATIVE ERROR OF THE FORM,
C
C          E(LOG(Y)) = -LOG( A + B * X ) ,
C
C      WHERE X IS THE INDEPENDENT VARIABLE, E(LOG(Y)) THE EXPECTED VALUE FOR
C      THE MODEL AND A AND B ARE PARAMETERS TO BE ESTIMATED FROM THE DATA.
C      THE LENGTH NP OF THE VECTOR OF PARAMETERS WHICH IS PASSED TO SUBROUTINE
C      SOLVE WILL VARY DEPENDING ON THE RESTRICTIONS PLACED ON THE MODEL.
C      IN THE LEAST RESTRICTIVE CASE WITH TWO PARAMETERS BEING ESTIMATED
C      FOR EACH OF THE CONSTITUENT DATA SETS (INDIVIDUAL A, INDIVIDUAL B),
C      THERE WILL BE NP = 2 * 4 = 8 PARAMETERS ESTIMATED IN A SINGLE CALL
C      TO SUBROUTINE SOLVE.  IN THE MOST RESTRICTIVE CASE WITH TWO PARAMETERS
C      COMMON TO ALL OF THE CONSTITUENT DATA SETS (COMMON A, COMMON B), THERE
C      WILL BE NP = 2 PARAMETERS TO BE ESTIMATED IN A SINGLE CALL TO
C      SUBROUTINE SOLVE.  OTHER RESTRICTIONS INCLUDE THE FITTING OF ONE OF
C      THE TWO PARAMETERS AS A COMMON PARAMETER FOR ALL DATA SETS WHILE
C      ALLOWING THE OTHER PARAMETER TO BE FITTED INDIVIDUALLY TO EACH OF
C      THE CONSTITUENT DATA SETS (NP = 5).  IN THE CASES WHERE INDIVIDUAL
C      PARAMETERS ARE FITTED TO SEPARATE DATA SETS IN A SINGLE CALL TO
C      SUBROUTINE SOLVE, SPECIAL CODE MUST BE INCLUDED WITH THE EVALUATION
C      SUBROUTINE (SUBROUTINE EVAL1 IN THIS EXAMPLE) TO ENSURE THAT THE
C      CORRECT PARAMETER ESTIMATES ARE USED IN THE CALCULATION OF FITTED
C      VALUES AND PARTIAL DERIVATIVES AT EACH DATA POINT.  THE USE OF AN
C      INDEXING ARRAY (ARRAY INDEX IN THIS EXAMPLE) TO ACHIEVE THIS RESULT
C      IS ILLUSTRATED IN THE FOLLOWING SECTION OF CODE.  INDEX IS DIMENSIONED
C      (166,2) WITH THE ROWS CORRESPONDING TO THE 166 DATA POINTS IN THE
C      COMBINED DATA SET AND THE COLUMNS CORRESPONDING TO THE TWO PARAMETERS
C      OF THE ASYMPTOTIC MODEL.  THE CONTENT OF INDEX(I,J), WHERE I CORRESPONDS
C      TO THE ITH DATA POINT, IS INITIALIZED IN THE SUBROUTINE SETUP TO AN
C      INTEGER BETWEEN 1 AND NP WHICH INDEXES THE LOCATION OF THE JTH
C      PARAMETER OF THE ASYMPTOTIC MODEL IN THE VECTOR OF PARAMETERS PASSED
C      TO SUBROUTINE SOLVE.  FOR THE RESTRICTIONS OUTLINED ABOVE, THE ARRAY
C      INDEX WOULD CONTAIN THE FOLLOWING INDICES AFTER INITIALIZATION -
C
```

```
C                 INDIVIDUAL A    INDIVIDUAL A     COMMON A       COMMON A
C     DATA POINT  INDIVIDUAL B     COMMON B     INDIVIDUAL B     COMMON B
C         1          1   5          1   5          1   2          1   2
C                    .   .          .   .          .   .          .   .
C                    .   .          .   .          .   .          .   .
C                    .   .          .   .          .   .          .   .
C         42         1   5          1   5          1   2          1   2
C
C         43         2   6          2   5          1   3          1   2
C                    .   .          .   .          .   .          .   .
C                    .   .          .   .          .   .          .   .
C                    .   .          .   .          .   .          .   .
C         83         2   6          2   5          1   3          1   2
C
C         84         3   7          3   5          1   4          1   2
C                    .   .          .   .          .   .          .   .
C                    .   .          .   .          .   .          .   .
C                    .   .          .   .          .   .          .   .
C         125        3   7          3   5          1   4          1   2
C
C         126        4   8          4   5          1   5          1   2
C                    .   .          .   .          .   .          .   .
C                    .   .          .   .          .   .          .   .
C                    .   .          .   .          .   .          .   .
C         166        4   8          4   5          1   5          1   2
C
C                  NP = 8         NP = 5         NP = 5         NP = 2
C
C     THE VECTOR OF PARAMETERS FOR THE LEAST RESTRICTIVE CASE WOULD
C     THEREFORE CONTAIN FOUR VALUES FOR PARAMETER A OF THE MODEL, FOLLOWED
C     BY FOUR VALUES FOR PARAMETER B OF THE MODEL, WITH BOTH SETS OF
C     PARAMETERS LISTED IN ORDER OF THE OCCURRENCE OF THE FOUR DATA SETS.
C     ON THE OTHER HAND, THE VECTOR OF PARAMETERS FOR THE MOST RESTRICTIVE
C     CASE WOULD SIMPLY CONTAIN AN ESTIMATE FOR PARAMETER A AND AN ESTIMATE
C     FOR PARAMETER B, IN THAT ORDER.
C
C     SUBROUTINES AND FUNCTIONS REQUIRED
C         USER-SUPPLIED       EVAL1,SETUP
C         NONLIN              SOLVE,DSPLAY
C         LINPACK             SQRDC,SQRSL,STRDI,SAXPY,SDOT,SSCAL,SSWAP,
C                             SNRM2,SCOPY
C         FORTRAN             FLOAT,ABS,MOD,AMAX1,MIN0,SQRT,SIGN,ALOG
C
C     HISTORY
C         25/06/82   PROGRAMMED (R.K. LOWRY)
C
C----------------------------------------------------------------------
C
      EXTERNAL EVAL1
C
      INTEGER JPVT(8),NITS,NPTS,NP,ND1,LUNIN,LUNOUT,INDEX,
     +        I,IFAIL,JOB,NDF,NDF1
      REAL Y(166),PARAM(8),PAR(8),VARC(8,8),DERIVS(1328),QRAUX(8),
     +     DIFF(166),WK1(36),ALPHA,RSS,X,ALOG,FLOAT,SS1,SS2,SS3,SS4,
     +     SS,VAR,VAR1,F
C
      COMMON X(166),INDEX(166,2)
      DATA ND1/8/,LUNIN/5/,LUNOUT/6/,NITS/30/,JOB/11000/
      DATA ALPHA/1.0E-6/
C
```

```
C     NOTE THAT THE TOTAL NUMBER OF POINTS IS DEFINED HERE WHILE THE NUMBER
C     OF POINTS IN EACH DATA SET IS DEFINED IN SUBROUTINE SETUP.
C
      DATA NPTS/166/
C
C     INITIAL PARAMETER ESTIMATES.
C
C     PARAMETER A.
      DATA PARAM(1)/3.739E-3/,PARAM(2)/3.462E-3/,PARAM(3)/1.883E-3/,
     +     PARAM(4)/1.809E-3/
C     PARAMETER B.
      DATA PARAM(5)/1.093E-4/,PARAM(6)/1.291E-4/,PARAM(7)/9.159E-5/,
     +     PARAM(8)/1.420E-4/
C
C     READ IN THE FOUR DATA SETS.
C
      DO 100 I=1,NPTS
         READ(LUNIN,10) X(I),Y(I)
         Y(I)=ALOG(Y(I))
  100 CONTINUE
C
C****************************************************
C
C     FIT THE FULL MODEL (INDIVIDUAL A AND B FOR EACH DATA SET).
C
      WRITE(LUNOUT,15)
      CALL SETUP(1,1,5,1)
      NP=8
C
      CALL SOLVE(Y,PARAM,ALPHA,NITS,NPTS,NP,ND1,LUNOUT,JOB,EVAL1,
     +           VARC,RSS,IFAIL,DERIVS,QRAUX,JPVT,DIFF,WK1)
      SS1=RSS
C
C****************************************************
C
C     FIT COMMON A, INDIVIDUAL B.
C
      WRITE(LUNOUT,20)
      CALL SETUP(1,0,2,1)
C
C     DERIVE THE INITIAL PARAMETER ESTIMATES FROM THE CONVERGED
C     PARAMETER ESTIMATES OBTAINED BY FITTING THE FULL MODEL
C     - MEAN OF 4 A'S AND INDIVIDUAL B'S
C
      PAR(1)=(PARAM(1)+PARAM(2)+PARAM(3)+PARAM(4))/4.0E0
      PAR(2)=PARAM(5)
      PAR(3)=PARAM(6)
      PAR(4)=PARAM(7)
      PAR(5)=PARAM(8)
      NP=5
C
      CALL SOLVE(Y,PAR,ALPHA,NITS,NPTS,NP,ND1,LUNOUT,JOB,EVAL1,
     +           VARC,RSS,IFAIL,DERIVS,QRAUX,JPVT,DIFF,WK1)
      SS2=RSS
C
C****************************************************
C
C     FIT INDIVIDUAL A'S AND COMMON B.
C
```

```
      WRITE(LUNOUT,25)
      CALL SETUP(1,1,5,0)
      NP=5
C
C     INITIAL PARAMETER ESTIMATES - INDIVIDUAL A'S AND MEAN OF
C     4 B'S.
C
      PAR(1)=PARAM(1)
      PAR(2)=PARAM(2)
      PAR(3)=PARAM(3)
      PAR(4)=PARAM(4)
      PAR(5)=(PARAM(5)+PARAM(6)+PARAM(7)+PARAM(8))/4.0E0
C
      CALL SOLVE(Y,PAR,ALPHA,NITS,NPTS,NP,ND1,LUNOUT,JOB,EVAL1,
     +          VARC,RSS,IFAIL,DERIVS,QRAUX,JPVT,DIFF,WK1)
      SS3=RSS
C
C************************************************
C
C     FIT COMMON A, COMMON B.
C
      WRITE(LUNOUT,30)
      CALL SETUP(1,0,2,0)
      NP=2
C
C     INITIAL PARAMETER ESTIMATES - MEAN OF 4 A'S AND MEAN OF 4 B'S.
C
      PAR(1)=(PARAM(1)+PARAM(2)+PARAM(3)+PARAM(4))/4.0E0
      PAR(2)=(PARAM(5)+PARAM(6)+PARAM(7)+PARAM(8))/4.0E0
C
      CALL SOLVE(Y,PAR,ALPHA,NITS,NPTS,NP,ND1,LUNOUT,JOB,EVAL1,
     +          VARC,RSS,IFAIL,DERIVS,QRAUX,JPVT,DIFF,WK1)
      SS4=RSS
C
C************************************************
C
C     CALCULATE STATISTICS FOR VARIOUS HYPOTHESES AND PRINT AN ANALYSIS
C     OF VARIANCE TABLE.
C
      WRITE(LUNOUT,35)
      NDF1=NPTS-8
      VAR1=SS1/FLOAT(NDF1)
C
C     COMMON A AND B.
C
      NDF=6
      SS=SS4-SS1
      VAR=SS/FLOAT(NDF)
      F=VAR/VAR1
      WRITE(LUNOUT,40) NDF,SS,VAR,F
C
C     COMMON A.
C
      NDF=3
      SS=SS2-SS1
      VAR=SS/FLOAT(NDF)
      F=VAR/VAR1
      WRITE(LUNOUT,45) NDF,SS,VAR,F
C
C     COMMON B.
C
```

```
      SS=SS3-SS1
      VAR=SS/FLOAT(NDF)
      F=VAR/VAR1
      WRITE(LUNOUT,50) NDF,SS,VAR,F
C
      WRITE(LUNOUT,55) NDF1,SS1,VAR1
C
      STOP
C
C   FORMAT STATEMENTS.
C
   10 FORMAT(F6.2,F7.2)
   15 FORMAT(///1H ,26HINDIVIDUAL A, INDIVIDUAL B)
   20 FORMAT(///1H ,22HCOMMON A, INDIVIDUAL B)
   25 FORMAT(///1H ,22HINDIVIDUAL A, COMMON B)
   30 FORMAT(///1H ,18HCOMMON A, COMMON B)
   35 FORMAT(///1H ,19HTESTS OF HYPOTHESES
     +          //1H ,36X,4HD.F.,6X,4HS.S.,6X,4HM.S.,9X,1HF
     +          /1H ,70(1H-))
   40 FORMAT(1H ,23HTEST FOR COMMON A AND B,7X,I10,2F10.4,F10.2)
   45 FORMAT(1H ,17HTEST FOR COMMON A,13X,I10,2F10.4,F10.2)
   50 FORMAT(1H ,17HTEST FOR COMMON B,13X,I10,2F10.4,F10.2)
   55 FORMAT(1H ,23HLEAST RESTRICTIVE MODEL,7X,I10,2F10.4
     +          /1H ,70(1H-))
      END




      SUBROUTINE SETUP(I1,IN1,I2,IN2)
C
C-------------------------------------------------------------------------
C
C     SUBROUTINE SETUP INITIALIZES THE ARRAY INDEX TO CONTAIN INDICES TO
C     A VECTOR OF PARAMETERS FOR EACH DATA POINT FOR ANY TWO PARAMETER
C     MODEL.
C
C     DESCRIPTION
C     THE INTEGER VALUES STORED IN THE ARRAY INDEX REFER TO THE LOCATION
C     OF PARAMETERS IN THE VECTOR OF PARAMETERS PASSED INTO SUBROUTINE
C     EVAL1 VIA SUBROUTINE SOLVE.  THE ROWS OF INDEX CORRESPOND TO THE
C     SEQUENCE OF X-VALUES DERIVED FROM THE COMBINATION OF A NUMBER
C     OF DATA SETS AND THE COLUMNS CORRESPOND TO THE ACTUAL PARAMETERS IN
C     THE ASYMPTOTIC MODEL FOR WHICH THE INDEXED PARAMETER IS TO BE
C     SUBSTITUTED DURING CALCULATIONS INVOLVING THE SPECIFIED DATA POINT.
C     THE NUMBER OF DATA SETS (NSETS) AND THE NUMBER OF DATA POINTS IN EACH
C     DATA SET (NUNITS) ARE DEFINED IN A DATA STATEMENT IN THE EXAMPLE CODE.
C
C     ON ENTRY
C       I1      INTEGER
C               THE LOCATION OF THE FIRST PARAMETER FOR THE MODEL IN THE
C               VECTOR OF PARAMETERS PASSED INTO SUBROUTINE SOLVE.
C       IN1     INTEGER
C               THE INCREMENT BY WHICH I1 SHOULD BE INCREASED FOR EACH NEW
C               DATA SET (USUALLY 0 OR 1).
C       I2      INTEGER
C               THE LOCATION OF THE SECOND PARAMETER FOR THE MODEL IN THE
C               VECTOR OF PARAMETERS.
C       IN2     INTEGER
C               THE INCREMENT BY WHICH I2 SHOULD BE INCREASED FOR EACH NEW
C               DATA SET (USUALLY 0 OR 1).
```

```
C
C      NOTE THAT THE VALUE ZERO PASSED IN AS A PARAMETER INCREMENT IS USED
C      WHEN THE PARAMETER IS COMMON TO ALL DATA SETS WHEREAS A VALUE OF
C      ONE IS USED WHEN THE PARAMETER IS TO BE ESTIMATED INDIVIDUALLY FOR
C      EACH DATA SET (PROVIDED THAT THE ESTIMATES FOR THIS PARAMETER ARE
C      STORED CONSECUTIVELY IN THE VECTOR PASSED TO SUBROUTINE SOLVE).
C
C      HISTORY
C          22/06/82  PROGRAMMED  (R.K. LOWRY)
C
C------------------------------------------------------------------------
C
       INTEGER NUNITS(4),INDEX,IP1,INC1,IP2,INC2,NSETS,I,J,K,N,
      +        I1,IN1,I2,IN2
       COMMON X(166),INDEX(166,2)
C
       DATA NUNITS(1)/42/,NUNITS(2)/41/,NUNITS(3)/42/,NUNITS(4)/41/
       DATA NSETS/4/
C
       IP1=I1
       INC1=IN1
       IP2=I2
       INC2=IN2
C
       N=0
       DO 100 I=1,NSETS
          K=NUNITS(I)
          DO 200 J=1,K
             N=N+1
             INDEX(N,1)=IP1
             INDEX(N,2)=IP2
  200     CONTINUE
          IP1=IP1+INC1
          IP2=IP2+INC2
  100 CONTINUE
       RETURN
       END



       SUBROUTINE EVAL1(P,I,ITASK,Z,IFAIL)
C
C------------------------------------------------------------------------
C
C      SUBROUTINE EVAL1 IS AN EXAMPLE OF A USER-SUPPLIED SUBROUTINE WHICH
C      ENABLES INDIVIDUAL PARAMETERS TO BE FITTED TO A NUMBER OF DATA SETS
C      IN ONE CALL TO SUBROUTINE SOLVE.
C
C      DESCRIPTION
C      THE ARRAY INDEX WHICH CONTAINS INDICES TO THE VECTOR OF PARAMETERS
C      PASSED TO SUBROUTINE SOLVE IS SUPPLIED TO SUBROUTINE EVAL1 IN A
C      COMMON BLOCK.  THESE INDICES ARE USED IN THE EVALUATION OF THE MODEL
C      FUNCTION AND FIRST-ORDER PARTIAL DERIVATIVES TO SELECT THE APPROPRIATE
C      PARAMETER ESTIMATES FOR EACH DATA POINT.  NOTE THAT THE EVALUATION
C      OF PARTIAL DERIVATIVES BY THIS METHOD REQUIRES AN INITIAL LOOP TO ZERO
C      ELEMENTS OF THE RESULTS VECTOR (ARRAY Z IN THIS EXAMPLE) THAT ARE
C      INAPPROPRIATE FOR THE PARTICULAR DATA POINT.  THE DESCRIPTIONS OF
C      THE ARGUMENTS OF. SUBROUTINE EVAL1 ARE IDENTICAL AND IN THE SAME
C      ORDER AS THOSE DETAILED FOR SUBROUTINE EVAL IN THE PREVIOUS
C      EXAMPLE (IN SUBROUTINE EVAL1, P IS USED IN PLACE OF PARAM AND Z IN
```

```
C        PLACE OF WK1 FOR SIMPLICITY).  AS THERE IS NO CALL TO SUBROUTINE
C        BATES IN THIS EXAMPLE, CODE FOR THE SECOND DERIVATIVES IS NOT
C        REQUIRED IN SUBROUTINE EVAL1.
C
C        HISTORY
C           22/06/82  PROGRAMMED  (R.K. LOWRY)
C
C-------------------------------------------------------------------
C
         INTEGER I,ITASK,IFAIL,I1,I2,INDEX,J
         REAL P(1),Z(8),X,ZERO,ONE,C,ALOG
         COMMON X(166),INDEX(166,2)
         DATA ZERO/0.0E0/,ONE/1.0E0/
C
         IFAIL=0
         I1=INDEX(I,1)
         I2=INDEX(I,2)
         IF(ITASK.NE.0) GO TO 1111
             Z(1)=-ALOG(P(I1)+P(I2)*X(I))
             RETURN
C
 1111 CONTINUE
         IF(ITASK.NE.1) GO TO 2222
             DO 100 J=1,8
                 Z(J)=ZERO
  100        CONTINUE
C
             C=P(I1)+P(I2)*X(I)
             Z(I1)=-ONE/C
             Z(I2)=-X(I)/C
             RETURN
C
 2222 CONTINUE
         IFAIL=1
         RETURN
         END



    INDIVIDUAL A, INDIVIDUAL B
```

INDIVIDUAL A, INDIVIDUAL B

PARAMETER ESTIMATES

RSS	1	2	3	4
.177814E+01	.373900E-02	.346200E-02	.188300E-02	.180900E-02
.174017E+01	.373892E-02	.388251E-02	.188256E-02	.195191E-02
.174014E+01	.373893E-02	.387048E-02	.188257E-02	.195661E-02
.174014E+01	.373893E-02	.387061E-02	.188257E-02	.195676E-02
.174014E+01	.373893E-02	.387061E-02	.188257E-02	.195677E-02
.174014E+01	.373893E-02	.387061E-02	.188257E-02	.195677E-02

5	6	7	8
.109300E-03	.129100E-03	.915900E-04	.142000E-03
.109324E-03	.117984E-03	.915922E-04	.138166E-03
.109323E-03	.118286E-03	.915921E-04	.138084E-03
.109323E-03	.118283E-03	.915921E-04	.138081E-03
.109323E-03	.118283E-03	.915921E-04	.138081E-03
.109323E-03	.118283E-03	.915921E-04	.138081E-03

PARAMETER ESTIMATES AT CONVERGENCE

PARAMETER	ESTIMATE	SE	T
1	.373893E-02	.338765E-03	11.04
2	.387061E-02	.341116E-03	11.35
3	.188257E-02	.240711E-03	7.82
4	.195677E-02	.282835E-03	6.92
5	.109323E-03	.615018E-05	17.78
6	.118283E-03	.648139E-05	18.25
7	.915921E-04	.402135E-05	22.78
8	.138081E-03	.593444E-05	23.27

COMMON A, INDIVIDUAL B

PARAMETER ESTIMATES

RSS	1	2	3	4	5
.372390E+01	.286222E-02	.109323E-03	.118283E-03	.915921E-04	.138081E-03
.218527E+01	.275297E-02	.124998E-03	.136717E-03	.785376E-04	.123132E-03
.217580E+01	.266170E-02	.127325E-03	.139388E-03	.806653E-04	.125519E-03
.217579E+01	.266080E-02	.127385E-03	.139464E-03	.806437E-04	.125508E-03
.217579E+01	.266058E-02	.127390E-03	.139469E-03	.806468E-04	.125512E-03
.217579E+01	.266058E-02	.127390E-03	.139469E-03	.806468E-04	.125512E-03
.217579E+01	.266058E-02	.127390E-03	.139469E-03	.806468E-04	.125512E-03

PARAMETER ESTIMATES AT CONVERGENCE

PARAMETER	ESTIMATE	SE	T
1	.266058E-02	.163752E-03	16.25
2	.127390E-03	.418179E-05	30.46
3	.139469E-03	.449172E-05	31.05
4	.806468E-04	.312702E-05	25.79
5	.125512E-03	.428531E-05	29.29

INDIVIDUAL A, COMMON B

PARAMETER ESTIMATES

RSS	1	2	3	4	5
.392484E+01	.373893E-02	.387061E-02	.188257E-02	.195677E-02	.114320E-03
.226716E+01	.360275E-02	.415897E-02	.733480E-03	.299352E-02	.112049E-03
.223496E+01	.372340E-02	.427524E-02	.989321E-03	.320669E-02	.109647E-03
.223489E+01	.372079E-02	.427400E-02	.981149E-03	.321265E-02	.109695E-03
.223489E+01	.372134E-02	.427451E-02	.981957E-03	.321326E-02	.109683E-03
.223489E+01	.372133E-02	.427451E-02	.981919E-03	.321327E-02	.109684E-03
.223489E+01	.372133E-02	.427451E-02	.981922E-03	.321328E-02	.109684E-03
.223489E+01	.372133E-02	.427451E-02	.981922E-03	.321328E-02	.109684E-03

PARAMETER ESTIMATES AT CONVERGENCE

PARAMETER	ESTIMATE	SE	T
1	.372133E-02	.227066E-03	16.39
2	.427451E-02	.230369E-03	18.56

```
3    .981922E-03  .195247E-03      5.03
4    .321328E-02  .207653E-03     15.47
5    .109684E-03  .300518E-05     36.50
```

COMMON A, COMMON B

```
        PARAMETER ESTIMATES
       RSS          1            2
.599467E+01  .286222E-02  .114320E-03
.590720E+01  .316998E-02  .106738E-03
.590719E+01  .316878E-02  .106803E-03
.590719E+01  .316876E-02  .106803E-03
.590719E+01  .316876E-02  .106803E-03
```

PARAMETER ESTIMATES AT CONVERGENCE

```
PARAMETER    ESTIMATE         SE            T
    1      .316876E-02  .270013E-03     11.74
    2      .106803E-03  .492846E-05     21.67
```

TESTS OF HYPOTHESES

	D.F.	S.S.	M.S.	F
TEST FOR COMMON A AND B	6	4.1670	.6945	63.06
TEST FOR COMMON A	3	.4356	.1452	13.19
TEST FOR COMMON B	3	.4948	.1649	14.97
LEAST RESTRICTIVE MODEL	158	1.7401	.0110	

5. References

American National Standards Institute (1966). *USA Standard Fortran X3.9-1966*, American National Standards Institute, New York.

Bates, D. M. and Watts, D. G. (1980). Relative Curvature Measures of Nonlinearity, *J. R. Statist. Soc., Ser. B 42*, 1-25.

Box, M. J. (1971). Bias in Nonlinear Estimation, *J. R. Statist. Soc., Ser. B 33*, 171-201.

Businger, P. and Golub, G. H. (1965). Linear Least Squares Solutions by Householder Transformations, *Numer. Math. 7*, 269-276.

Dongarra, J. J., Moler, C. B., Bunch, J. R., and Stewart, G. W. (1979). *LINPACK Users' Guide*, SIAM, Philadelphia.

Draper, N. R. and Smith, H. (1981). *Applied Regression Analysis*, 2nd ed., Wiley, New York.

Kennedy, W. J. and Gentle, J. E. (1980). *Statistical Computing*, Dekker, New York.

Ryder, B. G. and Hall, A. D. (1973). *The PFORT Verifier*, Computing Science Technical Report 12, Bell Laboratories, New Jersey.

Singleton, R. C. (1969). An Efficient Algorithm for Sorting with Minimal Storage, *Commun. ACM 12*, 185-187.

Solutions to Exercises

Chapter 1

Exercise 1.1 The Y-vector $(2.5, 1)$ may be represented on Fig. 1.2. The perpendicular from it to the solution locus yields $\theta \cong .5$. The solution locus is more curved here, and $\Delta\theta$ more unequally spaced, than at $\theta = 2.0537$, so both the intrinsic and parameter-effects nonlinearities will be greater than for the original data pairs in Table 1.1.

Exercise 1.2 The solution locus is a straight line, and the points corresponding to $\Delta\beta = 1$ are equally spaced.

Exercise 1.3 Since $\theta = \log \phi$, it follows that $(X_1^\theta, X_2^\theta) \equiv (X_1^{\log\phi}, X_2^{\log\phi})$ and the positions of the solution locus in Figs. 1.2a and b are identical.

Chapter 2

Exercise 2.7 $F(1, 1, .01) = 4052$; therefore, $S(\theta) = 11,889$. $\theta_U = 4.339$ and $\phi_U = 76.65$. There are no other values of θ for which $S(\theta) = 11,889$.

Chapter 3

Exercise 3.1

Locality	Parameter	Bleasdale-Nelder	Holliday	Farazdaghi-Harris	Asymptotic
Uraidla	α	.008897	.004143	.004516	.003871
	β	$.1861(10^{-3})$	$.1083(10^{-3})$	$.06892(10^{-3})$	$.1183(10^{-3})$
	θ, γ, or φ	.8646	$.06834(10^{-6})$	1.105	
	σ^2	.01056	.01058	.01054	.01035
Virginia	α	$.3149(10^{-4})$	.001664	$.3117(10^{-3})$	.001957
	β	$.05344(10^{-3})$	$.1499(10^{-3})$	$.3632(10^{-3})$	$.1381(10^{-3})$
	θ, γ, or φ	1.257	$-.08922(10^{-6})$	.8131	
	σ^2	.01410	.01446	.01409	.01420

Exercise 3.2

Parameter	Bleasdale-Nelder		Holliday		Farazdaghi-Harris		Asymptotic	
	U	V	U	V	U	V	U	V
α	161	1174	.3	.8	−4.4	−81.9	.01	.10
β	44	27	−.3	−.3	55	30	.03	.02
θ, γ, or φ	−.1	−.3	3.8	−4.2	.5	.4		
IN		.25	.22	.03	.04	.14	.12	.02	.03
PE		246.65	58.70	.07	.07	67.48	51.30	.03	.04

Exercise 3.3

	Test of:		
Locality	θ = 1	γ = 0	φ = 1
MG	−1.159	1.159	1.201
U	−.439	.389	.507
PL	−.599	.513	.652
V	1.354	−.540	−1.261

All t-values are not significant.

Exercise 3.4 No two simulation studies are identical because the data sets are generated randomly. The results should agree broadly with those in Fig. 3.2 and Table 3.3. The parameters giving rise to poorly behaved LS estimators are α and β in both the Bleasdale-Nelder and Farazdaghi-Harris models.

Chapter 4

Exercise 4.2 $\hat{\alpha}_{(4.6)} = \log \hat{\alpha}_{(4.1)}$, $\hat{\beta}_{(4.6)} = \exp \hat{\beta}_{(4.1)}$, $\hat{\gamma}_{(4.6)} = \exp [-\hat{\gamma}_{(4.1)}]$

Parameter	Pasture	Onion	Cucumber	Bean
α	4.417	6.584	1.935	3.114
β	3.400	12.18	2.156	8.218
γ	.9636	.6376	.6106	.6783

Exercise 4.4

Parameter	Pasture	Onion	Cucumber	Bean
α	2.763	1.201	2.427	8.027
β	.597	.113	1.035	.810
γ	.185	.178	1.194	.469

Chapter 5

Exercise 5.1 If $n = 12$, $F(3, 9; .05) = 3.86$, $1/(2\sqrt{F}) = .254$. The intrinsic curvatures of all seven data sets are less than this.

Exercise 5.2

Parameter	Dugong	Leaf	Wheat	Chem react.	Wheat	Potato	Rubber
α	.233	.887	3.941	.231	1.013	.117	.108
β	1.755	9.132	11.297	1.415	3.335	.329	2.375
γ	1.449	1.403	3.098	.074	1.131	.146	.771

Exercise 5.3 For dugongs, γ = exp (−.135248) = .8735

Bias $(\hat{\gamma})$ = .00195932(−.8735) + $\frac{1}{2}$(.000653123)(.8735) = −.001426

% Bias = $\frac{-.1426}{.8735}$ = −.163%

Similar calculations pertain to the other data sets.

Exercise 5.4

Data set	Intrinsic curvature	Parameter-effects curvature for model function	
		(5.4)	(5.7)
1	.243	.603	.536
2	.204	.531	.497
3	.228	.993	.951
4	.045	.198	.178
5	.178	1.370	1.084
6	.063	.555	.425
7	.141	.781	.658

Data set	Altered X-values, model function (5.4)			Altered X-values, model function (5.7)		
	α	β	γ	α	β	γ
1	.057	-10.736	1.289	.054	.684	-.150
2	.132	-.220	1.230	-.088	.543	-.020
3	.158	.012	4.266	.038	.433	-.184
4	.006	.023	.127	.001	.108	-.029
5	.106	.033	5.186	.023	.274	-1.976
6	.010	-.000	1.205	.001	.018	-.559
7	.011	.035	2.124	.004	.133	-.590

The results depend, of course, upon the set of X-values, but the reader should obtain results of the same order of magnitude.

Chapter 6

Exercise 6.3 $\hat{\theta}_1 = 15.6731$, $\hat{\theta}_2 = .999355$, $\hat{\theta}_3 = .0222197$, $\hat{\sigma}^2 = .0008552$.

Exercise 6.5

θ_1	.0338%
θ_2	.0011%
θ_3	.0004%
IN	.0002
PE	36.9821
$1/(2\sqrt{F})$	.271

Simulation studies should show close-to-linear behavior for each parameter despite the large value of PE. Correlation coefficients obtained using (2.18) should be very close to -1.

Exercise 6.6 The simulation study should show close-to-linear behavior for each parameter. This conforms with the indications given in Table 6.25; although the parameter-effects curvature .603 was greater than the critical value .304, it is still rather low in absolute terms.

Chapter 7

Exercise 7.1 α_1 = .0038638, α_2 = .0041469, α_3 = .0022630, α_4 = .0023141, β_1 = .00010369, β_2 = .00010997, β_3 = .000081341, β_4 = .00012819, γ = .58722(10^{-7}), RSS = .00019498.

Exercise 7.2

Description of fit or test	p	df	RSS	RMS
(C) Common α	5	161	2.176	
(B) Common β	5	161	2.235	
(A) Individual α and β	8	158	1.740	.01101

	df	Change in RSS	Mean square	F
(C) - (A): test of invariant α	3	.436	.145	13.2^{**}
(B) - (A): test of invariant β	3	.495	.165	15.0^{**}

Thus one must reject the hypothesis of either an invariant α or an invariant β.

Exercise 7.3   (a) $.59^{ns}$ (b) $.91^{ns}$ (c) $.31^{ns}$ (d) 7.20^{**}.

Exercise 7.4

Brown Imperial Spanish

	p	df	RSS	RMS
(G) Common β	3	80	.000115842	
(F) Common α	3	80	.000115240	
(E) Common α and β	2	81	.000126265	
(D) Individual α and β	4	79	.000114806	$1.453(10^{-6})$

	df	Change in RSS	Mean square	F
(E) - (D): test of invariant α and β	2	.000011459	$5.730(10^{-6})$	3.943^{*}
(F) - (D): test of invariant α	1	.000000434	.000000434	$.298^{ns}$
(G) - (D): test of invariant β	1	.000001036	.000001036	$.713^{ns}$

$$\alpha_1 = .00353925 \pm .000398200 \qquad \beta_1 = .000113563 \pm .00000528593$$

$$\alpha_2 = .00384017 \pm .000380698 \qquad \beta_2 = .000119745 \pm .00000506456$$

$$t_\alpha = \frac{.00384017 - .00353925}{.00055106} \qquad t_\beta = \frac{.000119745 - .000113563}{.0000073226}$$

$$= .546 \qquad\qquad = .844$$

$$t_\alpha^2 = .298 = F \qquad\qquad t_\beta^2 = .713 = F$$

White Imperial Spanish

	p	df	RSS	RMS
(G) Common β	3	80	.000145708	
(F) Common α	3	80	.000081581	
(E) Common α and β	2	81	.000375182	
(D) Individual α and β	4	79	.000081459	$1.031(10^{-6})$

	df	Change in RSS	Mean square	F
(E) − (D): test of invariant α and β	2	.000293723	.00014686	142.4^{**}
(F) − (D): test of invariant α	1	.000000122	.000000122	$.118^{ns}$
(G) − (D): test of invariant β	1	.000064249	.000064249	62.3^{**}

$$\alpha_1 = .00184907 \pm .000336171 \qquad \beta_1 = .0000925069 \pm .00000376126$$

$$\alpha_2 = .00200926 \pm .000323626 \qquad \beta_2 = .000137906 \pm .00000435093$$

$$t_\alpha = \frac{.00200926 - .00184907}{.00046674} \qquad t_\beta = \frac{.000137906 - .0000925069}{.0000057461}$$

$$= .343 \qquad\qquad\qquad = 7.90$$

$$t_\alpha^2 = .118 = F \qquad\qquad t_\beta^2 = 62.4 = F$$

Exercise 7.5

(a) 1975–1976: $\hat{\alpha} = .0221483$, $\hat{\beta} = .00204412$, Var $(\hat{\alpha}) = .715496(10^{-4})$,

Var $(\hat{\beta}) = .247915(10^{-8})$, RSS $= .00114381$

1976–1977: $\hat{\alpha} = .0219122$, $\hat{\beta} = .00245533$, Var $(\hat{\alpha}) = .257128(10^{-4})$,

Var $(\hat{\beta}) = .890933(10^{-9})$, RSS $= .000411051$

	p	df	RSS	RMS
(D) Common α and β	2	10	.0161376	
(C) Common α	3	9	.00155497	.00017277
(B) Common β	3	9	.0113063	
(A) Individual α and β	4	8	.00155486	.00019436

	df	Change in RSS	Mean square	F
(B) - (A): test of invariant β	1	.0097514	.0097514	50.2**
(C) - (A): test of invariant α	1	.00000011	.00000011	.00057ns
(D) - (C): test of invariant β, given invariant α	1	.0145826	.0145826	84.4**

(b) Alternative test of invariant α:

$$t^2 = \frac{(.0221483 - .0219122)^2}{(.715496 + .257128)10^{-4}} = .00057 = F$$

Alternative test of invariant β:

$$t^2 = \frac{(.00245533 - .00204412)^2}{(.890933 + 2.47915)10^{-9}} = 50.2 = F$$

For the fit of common α, $\hat{\beta}_1 = .00204453$, $\hat{\beta}_2 = .00245492$,

Var $(\hat{\beta}_1) = .124784(10^{-8})$, Var $(\hat{\beta}_2) = .124784(10^{-8})$,

RSS $= .00155497$.

Alternative test of invariant β, given invariant α:

$$t^2 = \frac{(.00245492 - .00204453)^2}{(.124784 + .124784)10^{-8}} = 67.5 \neq 84.4$$

(c) 1975-1976: $\hat{\alpha} = .0104398$, $\hat{\beta} = .00229835$, Var $(\hat{\alpha}) = .036459(10^{-4})$,

Var $(\hat{\beta}) = .0170076(10^{-6})$, RSS $= .0371831$

1976-1977: $\hat{\alpha} = .0162788$, $\hat{\beta} = .00255824$, Var $(\hat{\alpha}) = .681403(10^{-6})$,
Var $(\hat{\beta}) = .250772(10^{-8})$, RSS $= .00411696$

	p	df	RSS	RMS
(D) Common α and β	2	10	.147625	
(C) Common α	3	9	.0736451	
(B) Common β	3	9	.0570457	.0063384
(A) Individual α and β	4	8	.0413001	.0051625

	df	Change in RSS	Mean square	F
(B) − (A): test of invariant β	1	.0157456	.0157456	3.05^{ns}
(C) − (A): test of invariant α	1	.032345	.032345	6.27^{*}
(D) − (B): test of invariant α, given invariant β	1	.090579	.090579	14.3^{**}

(d) Alternative test of invariant α:

$$t^2 = \frac{(.0104398 - .0162788)^2}{.036459(10^{-4}) + .681403(10^{-6})} = 7.88 \neq 6.27$$

Alternative test of invariant β:

$$t^2 = \frac{(.00229835 - .00255824)^2}{(1.70076 + .250772)10^{-8}} = 3.46 \neq 3.05$$

Note: The result of the test of invariance of α in (a) contradicts that of the result of the same test in (c). To see whether model (7.2), the linear asymptotic yield-density model, is more appropriate than model (7.12), the logarithmic asymptotic yield-density model, for sweet lupin data, one requires replication at each density to estimate the variance of the yield Y, or functions of the yield such as log Y or 1/Y, to see which error assumption is the most appropriate. From data supplied by the Elliott Research Farm, it was found that Var (1/Y) was closer to being constant than Var (log Y) for 1975–1976, but the reverse was true for 1976–1977. More data are needed to settle the question, but the differing results in (a) and (c) serve to illustrate that different assumptions about the stochastic term can given rise to different inferences about the parameters.

Exercise 7.6

	p	df	RSS	RMS
(D) Common α	5	5	1949.05	
(C) Common β	5	5	1791.05	
(B) Common γ	5	5	409.194	
(A) Individual α, β, and γ	6	4	119.317	29.829

	df	Change in RSS	Mean square	F
(B) − (A): test of invariant γ	1	289.877	289.877	9.72*
(C) − (A): test of invariant β	1	1671.73	1671.73	56.0**
(D) − (A): test of invariant α	1	1829.73	1829.73	61.3**

Neither α, β, nor γ is invariant for these data in combination with model function (5.1).

Chapter 8

Exercise 8.1 $E(Y_t)$ must be of the form $1/f(\underset{\sim}{\theta}, X_t)$, where $f(\underset{\sim}{\theta}, X_t)$ is linear in the parameters $\underset{\sim}{\theta}$. Models (3.2) and (3.4), assuming additive error, are of this form and (6.6) may be converted to this form; if one parameter is held constant, models (3.1), (3.3), (4.8), and (6.5) are of this form.

Exercise 8.2 To derive (8.13), we start with (8.11) and expand it as follows:

$$S = \sum [Y(\alpha + \beta X + \gamma X^2) - 1]^2$$

$$= \sum [Y^2(\alpha + \beta X + \gamma X^2)^2 - 2Y(\alpha + \beta X + \gamma X^2) + 1]$$

$$= n + \sum \{(\alpha + \beta X + \gamma X^2)[Y^2(\alpha + \beta X + \gamma X^2) - 2Y]\}$$

$$= n + \alpha \sum [Y^2(\alpha + \beta X + \gamma X^2) - Y] - \alpha \sum Y$$

$$+ \beta \sum [XY^2(\alpha + \beta X + \gamma X^2) - XY] - \beta \sum XY$$

$$+ \gamma \sum [X^2Y^2(\alpha + \beta X + \gamma X^2) - X^2Y] - \gamma \sum X^2Y$$

To obtain the "normal" equations, (8.11) is differentiated with respect to α, β, and γ successively, and the resulting expressions are set equal to zero, yielding the following three relationships:

$$\sum [Y(\alpha + \beta X + \gamma X^2) - 1]Y = 0$$

$$\sum [Y(\alpha + \beta X + \gamma X^2) - 1]XY = 0$$

$$\sum [Y(\alpha + \beta X + \gamma X^2) - 1]X^2Y = 0$$

Substituting these relationships into the expanded form of S above, three of the seven terms vanish, yielding

$$S = n - \alpha \sum Y - \beta \sum XY - \gamma \sum X^2Y$$

The estimates are denoted α_0, β_0, and γ_0, respectively, and the residual sum of squares is denoted RSS:

$$RSS = n - \alpha_0 \sum Y - \beta_0 \sum XY - \gamma_0 \sum X^2Y$$

Expressions (8.20), (8.28), and (8.40) are derived in a similar fashion.

Exercise 8.3 $\gamma_{new} = \gamma$, $\alpha_{new} = 1/\alpha$, $\beta_{new} = \log(\alpha\beta)$.

Exercise 8.4   The second derivative of the right-hand side of (4.4) with respect to X is

$$\frac{(\alpha - \beta)\gamma\delta[(\delta - 1)(\gamma + X^\delta)X^{\delta-2} - 2\delta X^{2\delta-2}]}{(\gamma + X^\delta)^3}$$

Setting the numerator equal to zero and solving for X^δ yields

$$X^\delta = \frac{\gamma(\delta - 1)}{\delta + 1}$$

from which the result follows.

Exercise 8.5 The second derivative of the right-hand-side of (4.5) with respect to X is

$$\beta\gamma\delta \exp(-\gamma X^\delta)[(\delta - 1)X^{\delta-2} - \gamma\delta X^{2\delta-2}]$$

Setting the above equal to zero and solving for X^δ yields

$$X^\delta = \frac{\delta - 1}{\gamma\delta}$$

from which the result follows.

Exercise 8.6 We note that if γ is held constant, the model becomes linear. An adequate procedure might be to start with an estimate of γ, denoted γ_0, and to determine α_0 and β_0 by linear regression. The residual sum of squares is given by

$$RSS = \sum_{t=1}^{n} \left(\log Y_t - \alpha_0 + \frac{\beta_0}{X_t + \gamma_0} \right)^2$$

A series of trial values of γ_0 is tried and the set of estimates which gives the smallest RSS may be used as initial estimates with the Gauss-Newton algorithm.

Exercise 8.7 Consider model function (6.6),

$$E(Y) = \frac{X_1}{\phi_1 + \phi_2 X_1 + \phi_3 X_2}$$

which can be written as

$$E(Y) = \frac{1}{\phi_1/X_1 + \phi_2 + \phi_3 X_2/X_1}$$

The formula above is of the form required by expression (8.10) to produce a set of linear normal equations. Substituting the expression above into (8.10), one obtains (omitting the subscript t for simplicity)

$$S = \sum_{t=1}^{n} \left[Y \left(\frac{\phi_1}{X_1} + \phi_2 + \frac{\phi_3 X_2}{X_1} \right) - 1 \right]^2$$

Differentiating S with respect to ϕ_1, ϕ_2, and ϕ_3 successively, and setting the resulting expressions equal to zero, yields:

$$\phi_1 \sum \left(\frac{Y^2}{X_1^2} \right) + \phi_2 \sum \left(\frac{Y^2}{X_1} \right) + \phi_3 \sum \left(\frac{X_2 Y^2}{X_1^2} \right) = \sum \left(\frac{Y}{X_1} \right)$$

$$\phi_1 \sum \left(\frac{Y^2}{X_1} \right) + \phi_2 \sum Y^2 + \phi_3 \sum \left(\frac{X_2 Y^2}{X_1} \right) = \sum Y$$

$$\phi_1 \sum \left(\frac{X_2 Y^2}{X_1^2} \right) + \phi_2 \sum \left(\frac{X_2 Y^2}{X_1} \right) + \phi_3 \sum \left(\frac{X_2^2 Y^2}{X_1^2} \right) = \sum \left(\frac{X_2 Y}{X_1} \right)$$

The solutions ϕ_{10}, ϕ_{20}, ϕ_{30} to the set of three equations above may then be employed as initial estimates with the Gauss–Newton algorithm for model function (6.6). For model function (6.3), one may use $\theta_{10} = \phi_{20}/\phi_{10}$, $\theta_{20} = \phi_{30}/\phi_{10}$, and $\theta_{30} = 1/\phi_{20}$. For model function (6.5), $\phi_0 = \theta_{10}\theta_{30}$ may be used.

Exercise 8.8   The methods described in Sec. 8.4, appropriately modified, may be used.

Exercise 8.9 Because of the close relationship between model function (6.11) and model function (6.2), a method similar to that used for Exercise 8.6 may be developed. The same estimates may be used as initial estimates of the parameters of model function (6.1), with $\theta_1(\text{new}) = \exp(-\theta_1)$.

Chapter 9

<u>Exercise 9.1</u> $\hat{\phi}$ = 7.7966, Bias ($\hat{\phi}$) = -.006789, Var ($\hat{\phi}$) = 1.5075

$$F^T = F = \frac{1}{\hat{\phi}} (\log X) X^{\log \hat{\phi}} = .7716$$

$$H = - \frac{1}{\hat{\phi}^2} (\log X) X^{\log \hat{\phi}} (1 - \log X) = -.008284$$

Bias ($\hat{Y}$) = .7716(-.006789) + $\frac{1}{2}$(-.008284)(1.5075) = -.01148

Var ($\hat{Y}$) = (.7716)^2(1.5075) = .8975

<u>Exercise 9.2</u> For a linear model, Bias ($\hat{\theta}$) = 0; further, as F for a linear model is not a function of $\underline{\theta}$, it follows that H = 0. Thus, as both terms of (9.5) are zero, Bias ($\hat{Y}$) = 0.

<u>Exercise 9.3</u> Results for all densities and for all models are presented in the following table:

Density	Bleasdale-Nelder (3.1)	Holliday (3.2)	Farazdaghi-Harris (3.3)
21	.003956	.0004239	.002659
80	.0009687	.0001299	.0006238
155	-.002297	.0001695	-.001523

The relative magnitude of the biases conform to what might be expected from the curvature measures of nonlinearity presented in Table 3.1. At each density, the Bleasdale-Nelder model exhibits the greatest bias, the Holliday model exhibits the least bias, and the Farazdaghi-Harris model occupies an intermediate position.

<u>Exercise 9.4</u>   The parameter-effects nonlinearity measure indicates much greater nonlinearity in model function (4.4) than in model function (4.14).   Since the correlation matrices are the same, parameter correlations do not assist in predicting the estimation behavior of model functions.

Author Index

Subject Index